# GROUND WORK

# GROUND

~ *Conservation in American Culture* ~

# WORK

*Char Miller*

FOREST HISTORY SOCIETY

The Forest History Society is a nonprofit, educational institution dedicated to the advancement of historical understanding of human interaction with the forest environment. The Society was established in 1946. Interpretations and conclusions in FHS publications are those of the authors; the Society takes responsibility for the selection of topics, the competence of the authors, and their freedom of inquiry.

**THE FOREST HISTORY SOCIETY**

This book is published with support from the Lynn W. Day Foundation for Forest History Publications.

Printed in the United States of America

Forest History Society
701 William Vickers Avenue
Durham, North Carolina 27701
(919) 682-9319
www.foresthistory.org

First edition

Design by Zubigraphics, Inc.

Chapters in this book are adapted and reprinted from other sources with the kind permission of the original publishers; please see page 156.

Library of Congress Cataloging-in-Publication Data

Miller, Char, 1951-
  Ground work : conservation in American culture / Char Miller.
      p. cm.
  Includes bibliographical references.
  ISBN 978-0-89030-069-5 (hardcover : alk. paper)
  1.  Forest conservation--United States--History. 2.  Forests and forestry--United States--History.  I. Title.
  SD412.M55 2007
  333.75'160973--dc22
                        2007001833

*For Edgar B. Brannon*

# *Contents*

❦

A Greater Good: An Introduction . . . . . . . . . . . . . . . . . . . . . . . . . . . . . . . . 1

**I. New Perspectives**

1. Pivot Point . . . . . . . . . . . . . . . . . . . . . . . . . . . . . . . . . . . . . . . . . . . . . . . 11
2. Why *Garden and Forest* Mattered . . . . . . . . . . . . . . . . . . . . . . . . . . . 21
3. Wooden Politics . . . . . . . . . . . . . . . . . . . . . . . . . . . . . . . . . . . . . . . . . . . 26
4. What Really Happened in the Rainier Grand Hotel? . . . . . . . . . . . . . . 39
5. Sawdust Memories . . . . . . . . . . . . . . . . . . . . . . . . . . . . . . . . . . . . . . . . 50

**II. Survey Lines**

6. Eminent Domain . . . . . . . . . . . . . . . . . . . . . . . . . . . . . . . . . . . . . . . . . . 61
7. Groves of Academe . . . . . . . . . . . . . . . . . . . . . . . . . . . . . . . . . . . . . . . . 77
8. Grazing Arizona . . . . . . . . . . . . . . . . . . . . . . . . . . . . . . . . . . . . . . . . . . 87
9. Back to the Garden . . . . . . . . . . . . . . . . . . . . . . . . . . . . . . . . . . . . . . . . 96
10. Logjam . . . . . . . . . . . . . . . . . . . . . . . . . . . . . . . . . . . . . . . . . . . . . . . . . 110

**III. Scene Change**

11. Snapshot, 1897 . . . . . . . . . . . . . . . . . . . . . . . . . . . . . . . . . . . . . . . . . . 117
12. Green Screen . . . . . . . . . . . . . . . . . . . . . . . . . . . . . . . . . . . . . . . . . . . 121
13. Trees, Sprawl, and Urban Politics . . . . . . . . . . . . . . . . . . . . . . . . . . . 138

Afterword: An Open Field . . . . . . . . . . . . . . . . . . . . . . . . . . . . . . . . . . . . . 149

Acknowledgments . . . . . . . . . . . . . . . . . . . . . . . . . . . . . . . . . . . . . . . . . . . 155

Endnotes . . . . . . . . . . . . . . . . . . . . . . . . . . . . . . . . . . . . . . . . . . . . . . . . . . 159

Index . . . . . . . . . . . . . . . . . . . . . . . . . . . . . . . . . . . . . . . . . . . . . . . . . . . . . 171

About the Author . . . . . . . . . . . . . . . . . . . . . . . . . . . . . . . . . . . . . . . . . . . 177

# A Greater Good: An Introduction

The title was something of a snore, but the placement of "The Relation of Geography to Timber Supply," William B. Greeley's 1925 contribution to *Economic Geography*, could not have been better: it was the lead article in the first number of the inaugural issue.

Its editor had invited Greeley, as the chief of the USDA Forest Service, to submit a paper that would help the new periodical meet the growing need for "a full knowledge of the natural resources of the world, and a better understanding of the natural conditions to which man must more carefully adapt himself as population increases and the burden upon the land is made heavier." Greeley complied, introducing his subject with an economic calculation that he believed defined Americans' relationship to their forested estate. A nation's source of wood products is "largely governed by the cost of growing timber at home as compared with the cost of hauling it from the nearest virgin forests still available for exploitation." In this context, enacting forest conservation would be difficult, for in the "long run, forestry is pitted against transportation."[1]

There was nothing new in this tension. Because most "of the industrially aggressive nations have lived in forested regions, and most have been liberal users of timber," they have followed a three-stage pattern of consumption that the United States was beginning to replicate. "At first they have cut freely from their own virgin forests as long as the supply lasted," and when intense harvesting exhausted supplies, that was when they were compelled to seek new supplies elsewhere, through trade and barter. Finally, these societies "have settled down to the systematic growing of wood on all the land that could be spared for the purpose, still finding it necessary or convenient in many cases to import a substantial part of their national requirements from other countries whose virgin forests have not yet become depleted or whose timber culture produces an exportable surplus." Entering the industrial revolution later than some of its European counterparts, and blessed with an astonishingly thick forest cover, the United States of the mid-1920s remained in the first stage of development: "By far the greater part of the wood we use is still obtained from our virgin forests."[2]

That situation was about to change, and swiftly so, Greeley predicted, a result of the United States' insatiable and gargantuan appetite for wood fiber. "We use annually about 12 billion cubic feet of saw-log timber, or nearly half of the quantity consumed in the world. Our use of all forest products, including pulpwood, road ties, mine timbers, and fuel wood, aggregates 22 ½ billion cubic feet, or about two-fifths of the yearly consumption in the entire world." Those staggering rates were linked in "intimate relation" to the rise in living standards and the economy's robust growth since the late nineteenth century, making "our problem far more serious" that that of other nations. "We must find, almost overnight, a fresh source of raw material sufficient to supply 60 or 70 million tons of forest products annually. Instead of a gradual industrial evolution, the change is coming with the suddenness of an economic crisis."[3]

Nothing more graphically illustrated his point than three accompanying Forest Service maps of the United States depicting the evolution of "Area of Virgin Forest" in 1620, 1850, and 1920. When the English settlers made landfall in Massachusetts, they faced what one of them described as a "howling wilderness," containing, Greeley estimated, 820 million acres of uncut timber stretching from Maine to East Texas, with massive quantities in the interior-mountain West and Pacific Northwest. Although the 1620 map, with all its black ink, suggested that these forests were as dense as they were untouched by human hand, the first Europeans knew better, declaiming over the parklike quality of landscapes that contained elk and other large browsers not found in thick woods; natural fire cycles and Native Americans' use of fire to sustain grazing lands had opened up these forests. Yet by any measure, one of the most striking environmental characteristics of the New World—and what made it *new*—was its seemingly infinite wooded expanse.

By 1850, that was still (largely) true. However much forest cover had been logged, girdled, burned, or grazed over, intense human manipulation had been limited to eastern and southern river drainage systems. It was along those waterways that agriculture was most possible, and their flow to the sea was what propelled an endless stream of log rafts to market in river town and coastal port. Take Maine: "There were in 1837, as I read, two hundred and fifty sawmills on the Penobscot and its tributaries above Bangor," Henry David Thoreau recounted in *The Maine Woods* (1848), "and they sawed two hundred millions of feet of boards annually." When that harvest was tallied up with the logs rolled into the snowmelt surge pushing down the "the Kennebec, Adroscoggin, Saco, Passamaquoddy, and other streams," it is "no wonder that we hear so often of vessels which are becalmed off our coast being surrounded for a week at a time by floating timber." Northern Maine, however, was essentially unscathed, as were the Adirondacks, Appalachians and Ozarks, the Great Lakes, the Mississippi River and Ohio River bottomlands, and the far-distant western terrain. "The first 230 years of settlement and industrial expansion," Greeley concluded, "had made relatively slight inroads."[4]

The subsequent seven decades told a far different tale. With the unleashing of the industrial revolution, whose most impressive symbol was the wood-hungry railroad, once-wild lands were tied to the national and global marketplace. With new, more efficient tools of production, greater capitalization, and complex corporate structures, lumbering "ceased to be a village industry," Greeley observed. Becoming a "big business," it reached out with "unequaled driving force in manufacture and merchandizing," teaching "the American people to use wood in unheard-of quantities." Before the Civil War, "the per capita consumption of wood probably did not exceed 100 board feet. By 1906 it had become 506 board feet," and the population had increased from 26 million in 1860 to 98 million in 1910. Against this onslaught, neither hardwood nor softwood stood a chance.

At the time when Greeley wrote, the southern pineries, once covering upwards of 130 million acres, were playing out; their peak production had been recorded in 1916, and the "last great migration of American sawmills [was] underway—across the Great Plains to the virgin forests of the Pacific Coast." With this transition came soaring costs in the transportation of timber, and a spike in costs to consumers, leading to a doubling of lumber prices between 1913 and 1925. They would continue to rise, for the "true measure of timber supply is not quantity but availability": with cheap timber three thousand miles west of the population centers of the United States, and with no diminution of the nation's wood hunger, he wrote, "well-nigh famine prices" would remain the norm.[5]

That bad economic news was, paradoxically, good for foresters and conservationists. Until the mid-1920s, "forest conservation" was practiced not on private land but on public, on the many national forests under Chief Greeley's purview. Once the "illusion of inexhaustible virgin forests had spent itself," he wrote, the idea of pursuing more responsible land management strategies could begin to percolate "down into the counting house and the directors' board room." It had better. As the Europeans had discovered earlier, "no solution of our forest problem is possible without the generous growing of trees," which meant that forestry was "the only way to re-establish an adequate supply of timber in the United States."[6]

Given his readers at *Economic Geography* and his proclivities—Greeley was arguably the most business-oriented of the Forest Service chiefs and, upon his resignation from the agency in 1928, became executive secretary of the West Coast Lumberman's Association—his tight focus on the link between economics, geography, and forestry is understandable. His emphasis, however, is not an entirely accurate reflection of how the idea of conservation emerged first in Europe, or how and why ultimately it crossed the Atlantic, where by turns it was resisted and embraced.

More a product of European state formation, forest conservation grew in tandem with the development of powerful governmental authority in the late eighteenth century, well before industrialization dominated Old World economies. The historical development of scientific forestry, James C. Scott notes, "cannot be understood outside the larger context of the centralized state-making initiatives" of early modern Europe; "the new forestry science was a subdiscipline of what was called cameral science, an effort to reduce the fiscal management of a kingdom to scientific principles that would allow systematic planning."[7]

The trans-Atlantic migration of the concept of state paternalism in land management also commenced before the economy took hold in the United States. Historian John Reiger has demonstrated that beginning in the antebellum era, voluntary associations of hunters and anglers advocated local controls to preserve dwindling wooded habitat and riparian ecosystems, "believing that if wildlife, forests, and other natural resources were to be saved, concepts and practices already developed in aristocratic societies on the other side of the Atlantic would have to be adopted." These early conservationists were critical to framing the post-Civil War debates over the creation of the first national park at Yellowstone. They were also essential to the intensifying press coverage of conservation, itself a consequence of the publication of George Perkins Marsh's *Man and Nature* (1864) and the launch of *Field & Stream* (1873) and *Garden and Forest* (1888). When in response to this heightened publicity, Congress established in 1881 a division of forestry in the Department of Agriculture, its chronic underfunding was in some sense less important than its mere existence. As Bernhard Fernow, its third director, recognized, the agency was a beachhead from which to extend the notion of a federal regulatory regime, similar in reach and scope to that of his native Germany.[8]

Such expansion would not occur on his watch: it was not until 1905 that President Theodore Roosevelt created a full-fledged Forest Service with statutory power to manage the national forests and named the European-trained forester, Gifford Pinchot, then head of the Bureau of Forestry, to be its first chief. As the arc of Pinchot's career in Washington suggests, the ambitions he shared with Fernow were harder to realize on the ground than they were to imagine. Deciding which lands would be managed and under what conditions; countering stiff opposition determined to derail regulation and management of the public domain; grappling with allies, among them John Muir, over the use and preservation of the woods; and later, defending the agency's mission against what he perceived to be its internal detractors, among them William B. Greeley—all this and more revealed the stony soil in which conservation sought to sink its roots.

The success of these initial conservationist ambitions depended, as all new ideas do, on the ability of their advocates to educate the rising generation. Conservationism was eminently teachable, but late-nineteenth-century America had no schools whose curricula included explicit instruction in conservation or forestry. That would begin to change in the final years of the 1890s, at institutions small and large. One of the more unusual experiments occurred at the University of the South, in Sewanee, Tennessee. Led by its innovative vice chancellor, Benjamin L. Wiggins, in 1900 the school placed its nearly 10,000-acre campus on the logged-over Cumberland Plateau under sustained management from the Bureau of Forestry; the contract with the Pinchot-led agency enabled the school slowly to shift the basis of resource use on what it called the Domain and begin to regenerate forest cover. Conservation was the means to a new South.

Wiggins was also convinced that the teaching of forestry would revolutionize academic education, and in this he was not wrong, though his university was unable to generate the financing to start a forestry program. Others had the resources: in 1898 Fernow had left public service to build a forestry school at Cornell; Gifford Pinchot and his family endowed a graduate school at Yale; and Carl Schenck, another German forester, then working on George Vanderbilt's estate in North Carolina, ran a nondegree-granting vocational program. The tensions between these ambitious men, like the differences in their educational philosophies, are indicative of the manifold possibilities then awaiting their charges in the new profession. Of the three programs, only Yale's would survive, but that the Cornell and Biltmore schools closed their doors should not diminish the impact that their birth had on a culture that only recently had come to learn of conservation or forestry.

Placing those ideas into action produced a different form of education. So Gifford Pinchot learned when he traveled throughout Arizona at the beginning of the twentieth century and discovered, among other things, that in the Southwest, grazing, not timber management, would be what he called the "bloody angle," the issue of greatest initial controversy on the national forests. Coming to terms with the world of wool forced him to rethink the agency's organizational structure, an adaptation to local conditions and unforeseen pressures that had enduring significance. The first chief's nimble response was one of the attributes its thirteenth chief, Jack Ward Thomas, lauded in a speech to his leadership team in the early 1990s; the agency's past, he felt, continued to matter to its evolving present.

A legacy of social consciousness also runs through the history of conservation and forestry. Chief Greeley had alluded to this commitment to the disadvantaged when, in the final words of his *Economic Geography* article, he proposed that one ramification of a national commitment to conservative forest management was the maintenance of a "vigorous rural population"; its economic viability and community sustainability were at one with the agency's mission. Not all foresters—private or public—shared his perspective, but from the profession's beginning in the United States, some of its adherents had pressed the case that conservationism had cultural clout precisely because it was concerned with elevating those who lived on the margins. As he wrote up a land management scheme for George Vanderbilt's Biltmore Estate, Gifford Pinchot sketched out a forest plan for the Cherokee of western North Carolina. His certainty that conservative timber harvesting on their lands would net a steady stream of income, and that these funds would grant them a political freedom and social status they otherwise might not attain, resurfaced in the 1930s. During the Great Depression, John Collier, commissioner of Indian Affairs, proposed a New Deal in Indian forestry, believing that a reconsolidation of communal (tribal) property ownership would open the way for the implementation of controlled timber management; this would "bring to the Indians power to manage their own affairs and the self-respect which such power insures."[9] Seventy years later, the claim that forestry can empower the dispossessed had gone global: the Forest Stewardship Council, among other international groups focused on forest product certification, has stressed as one of its central principles that "green forestry" must be a socially ameliorative force for the people of tropical rainforests.

To engage the commonweal requires conservationists who understand one another, a particularly difficult task since World War II. If the debate among foresters, conservationists, and preservationists manifest in the Progressive Era seemed fraught, it paled in comparison with the ideological tensions in the 1960s that fractured their fragile alliance. As the environment achieved a newfound political cachet with the publication of Rachel Carson's lightning-rod book, *Silent Spring* (1962), and the enactment of landmark legislation, such as the Wilderness Act (1964), the National Environmental Policy Act (1970), and the Clean Air and Clean Water acts of the early 1970s, it opened the possibility for ideological divergence. The politics of the environment became hotly contested and not infrequently brutal. The Seventh American Forest Congress (1996) was called in the hopes of muting the cacophony, and to an extent, it was successful in defining a way by which the environment's many voices could learn a common language that described a viable middle ground.

A single conference can only do so much, however. In this age of visual media, more powerful (and often conflicting) representations of nature, and the animating vision we hold of it, have appeared in environmental documentaries. This medium, for all its evocative appeal, imaginative stimulation, and narrative force, is not free from the culture in which it is produced. When in the 1920s, filmmakers Robert Flaherty and John Grierson took their cameras into the wild, their foci were shaped by their political leanings and professional aspirations. Flaherty's *Nanook of the North* (1922) sought to capture the exotic, an Arctic people so unlike and inexplicable to his audience; Grierson's *The Drifters* (1929), a study of North Atlantic fishermen, hoped to explain why these men's economic situation was akin to British urban labor. Seeking an equilibrium between aesthetics and politics ever since has framed the nature documentary, from Pare Lorenz to Jacques Cousteau to Walt Disney, reminding us just how tricky such balancing acts can be.

None of these are more complex than the contemporary struggle to live lightly on the land. It has become almost impossible in the United States of early twenty-first century. That is not because we care so little for forest-draped mountains, wild-grass prairies, or crystal-clear streams; as with other generations, we pledge allegiance to America the Beautiful. But then we construct a built environment that paves over the very landscape we so praise. San Antonio is typical of this developmental pattern and its concrete consequences. Its sprawling form—1.3 million citizens living in a fully platted county of more than 1,200 square miles—has engulfed the rolling, oak-studded hills to the north and west, under which lies its sole-source aquifer. A widening arc of freeways meets the automotive needs of those seeking new housing on the expanding crabgrass frontier. Using trees as their marker species, environmental activists have battled for a legislative brake in the guise of a tree preservation ordinance, only to watch as developers gutted its provisions with the aid of a compliant city staff and council.

Urban reforestation, often touted as an antidote to significant loss of tree canopy and the creation of green cities, is tough to imagine in San Antonio, committed as its commercial elite and power brokers are to the well-bulldozed landscape. Grim too are the implications this one city's unsettling record holds for the other swelling cities of the Southwest—Houston, Dallas, Denver, Salt Lake, Las Vegas, Phoenix, Los Angeles, San Diego. Because these metropolises are in the national driver's seat—seven of the country's nine largest cities are located in the region—the experience of San Antonio and its urban peers bodes ill for the new century.

I hope I am wrong, not least because I live in San Antonio and have grown to love its lifeways. Recognizing how better to inhabit and think about this semiarid landscape has come from residing in the Alamo City for more than two decades. Over those years, too, came a shift in my academic interests, from intellectual and cultural history to urban and environmental. For that transformation I can thank historian Richard White. In 1985, I read his extended analysis of the central questions guiding environmental historiography in the *Pacific Historical Review* and found in his rigorous analysis of its scholarly claims and enthusiasms a dazzling array of possibilities for how to capture the complicated interplay between people and place. I have been entranced ever since.

**Divider photo:** *When not on loan for exhibition, Sanford Gifford's "Hunter Mountain, Twilight" (1866) was prominently displayed by James Pinchot in his home. Pinchot's collection of Hudson River School images, many of which showed cutover land, had a notable impact on his son Gifford's thinking. For more about this, see Chapter Eleven.* (USDA Forest Service)

CHAPTER ONE

❧

# *Pivot Point*

In 1876, German forester Bernhard E. Fernow left his native land for the United States in search of work. But his professional ambition dovetailed neatly with a romantic impulse. Several years earlier, he had fallen in love with Olivia Reynolds, a young American who had been living in Gottingen; they became engaged, and when she left for home in 1875, Fernow vowed to follow her as soon as possible. The 1876 U.S. Centennial Exposition, to be held in Philadelphia, was his ticket west. He secured appointment as an official Prussian observer to the celebration of American Independence, visited the great exhibition, and most likely attended the concurrent sessions of the second annual meeting of the American Forestry Association. But his heart lay elsewhere; he had come a long way to court his future wife.

Their marriage thereafter opened the way for him as a new citizen to contribute significantly to the establishment of his profession in the United States, a close relationship between family and state that his wife happily took credit for: "If anyone should ask me who was the originator of the forestry movement in this country, I should modestly reply, 'It was I.'" However tongue-in-cheek, her remark makes it clear that the start of Fernow's career was decidedly different from that of those Americans who would enter forestry's ranks in the late nineteenth century. But it is equally important to note that his experience neatly embodied the migration of European ideas that beginning in the 1870s did so much to introduce forestry to the American mind and landscape.[10]

That decade marks the beginning of a remarkable period during which the United States proved particularly receptive to a host of European notions about the proper role of the government in setting national policy. Until the early 1940s, the "reconstruction of American social politics" was intimately bound up with "movements of politics and ideas throughout the North Atlantic world that trade and capitalism had tied together," historian Daniel T. Rodgers has argued:

*This was not an abstract realization, slumbering in the recesses of consciousness. Tap into the debates that swirled throughout the United States and industrialized Europe over the problems and miseries of "great city" life, the insecurities of wage work, the social backwardness of the countryside, or the instabilities of the market itself, and one finds oneself pulled into an intense, transnational traffic in reform ideas, policies, and legislative devices.*

This "Atlantic era in social politics" required the creation of a "new set of institutional connections" with European societies, "new brokers" who would facilitate the intellectual exchange, and a cultural shift that would allow Americans to suspend, perhaps for the first (and only) time, their "confidence in the peculiar dispensation of the United States from the fate of other nations."[11]

In his compelling analysis, Rodgers makes only the briefest mention of forestry, and it is a late reference at that: in 1936 "a delegation of public foresters, including the chief of the U.S. Forest Service himself," were among those whose travel to Germany the Oberlaender Trust underwrote, funneling these "social policy experts through the familiar stations of German social progress." They may have been the last of such delegations. Yet in the preceding sixty years, the number of formal and informal exchanges had been continuous and of crucial importance to the development of the American forestry profession, none more so than the arrival in the United States in 1876 of that first essential broker, Bernhard Fernow.[12]

He was not the only participant in this rich trans-Atlantic cultural process, and given that Rodgers's model depends on an *American* hunger for European knowledge, perhaps he is not the best marker of it; he brought his technical knowledge of forestry with him. Other Americans of the 1870s who were concerned about the rapid devastation of the nation's wooded estate, and who believed that European scientific forestry might correct this difficult problem, had to seek out the relevant information. They avidly read British and German forestry texts, corresponded with their authors, joined the recently formed American Forestry Association (1875) or similar organizations that gave them access to additional information on European forestry, and traveled abroad to learn directly from European foresters. In so doing, they testified to the power of ideas to change social policy and alter political behavior.

For this cohort, there was much that needed changing. As early beneficiaries of the industrial revolution, they were also acutely aware of its remarkable and unrestrained consumption of wood. To generate the steam necessary to power industrial machinery and the new transportation mechanisms, as well as to construct massive new cities and shore up innumerable mines, required the clearcutting of once-bountiful forests; wholesale harvesting spread from the Great Lakes to the South and later to the Rocky Mountains and Pacific coast. As the trees crashed down, alarmed voices rose up. The *Sacramento Daily Union* warned in 1878 that if the then-current rate of logging continued unchecked, "the exhaustion of the forest growth in the Sierra is only a question of some ten years, and…if the rate of consumption is increased, the catastrophe will occur considerably sooner." Similarly vexed was Interior Secretary Carl Schurz, who a year earlier had predicted that it was only a matter of time before the nation's dire shortage of wood would jeopardize its capacity to build new homes. A timber famine seemed at hand.[13]

Even as observers fretted over this dangerous possibility and projected a catastrophe that would rival the fall of the ancient civilizations of the Mediterranean—they compared the future of the United States to a treeless (and impotent) Greece—these images of doom led some to ponder how best to change the status quo. One of those was Franklin B. Hough, a physician and statistician who understood forestry to be "a composite of natural history, geology, mathematics, and physics." From his New York State home, Hough had been assessing Census reports that reflected what those involved in the lumber business already knew: extensive harvests had depleted most eastern and Great Lakes forests, and southern and western woods were now under assault. This economic fact required a social prescription, Hough believed, and in 1873 he presented his findings in a paper entitled "On the Duty of Governments in the Preservation of Forests" to the American Association for the Advancement of Science. That governments had such an obligation was a radical departure for a nation wedded to laissez-faire capitalism. But Hough's reading in the scientific literature, his visits to Europe, and his extensive correspondence with German forester Dietrich Brandis offered him access to different models of governance; these allowed him to suggest alternatives to the rapid transfer of public wooded lands to corporate interests or homesteaders.[14]

The text he most depended on was George Perkins Marsh's seminal volume, *Man and Nature: The Earth as Modified by Human Action* (1864),

**Franklin B. Hough (left) and George Perkins Marsh looked to European scientific forestry for help with American forest problems.**

which the American diplomat had written while living in Italy. Marsh (and by extension, Hough) was convinced that deforestation of the European landscape had been responsible for the decline and fall of its great civilizations; he was equally certain that the only way to reverse this process—which he and Hough feared the United States was in danger of replicating—was to understand the close link between human profligacy and woodland devastation. To restrain human consumption of natural resources, to abate "the restless love of change which characterizes us," Marsh proposed the adoption of a conservative land management policy to be administered through a paternal form of government reminiscent of France and Germany, the leaders in European forestry.[15]

The wide gap between their leadership and American lethargy was what would impel Hough and other contemporaries periodically to petition Congress and various presidents to enact social legislation to protect the citizenry and enhance the commonwealth. For some reformers, Rodgers notes, this led to calls for top-down initiatives focused on better housing, sewage treatment, or efficient transportation. Forest advocates sought legislation to reflect their faith in the duty and capacity of government to regulate resource exploitation. Hough was an assiduous lobbyist and prolific publicist, and his campaigning bore fruit in 1876, when he was appointed to report on the nation's "forest supplies and conditions."

The fact of his 650-page document is as critical as the conclusions he reached within it; never before had the federal government published such an in-depth finding on the subject. What Hough found was troubling, but then he expected to be disturbed by the lumber industry's terrible and swift cutting of timber and the resulting scarred and battered terrain; his prior studies had paved the way for this negative reaction.[16]

Another who partly shared Hough's worries was Charles Sprague Sargent, head of the Arnold Arboretum and future publisher of *Garden and Forest*, which would become the central forum in the 1880s for sustained discussions about forestry and conservation in the United States. Encouraged by Interior Secretary Carl Schurz, who wanted a detailed analysis for the 1880 census on the nation's available timber supplies, the Smithsonian Institution hired Sargent to write what would become the *Report on the Forests of North America*; it appeared in 1884. Based on Sargent's on-site investigations, it delineated the distribution of tree species and forest densities and assessed the economic value of the nation's varied woods. Sargent concurred with Hough's assessment that there had been rapid harvesting of timber in New England, the mid-Atlantic, and the Midwest, but it was also true that the vast forests of the South and Pacific Northwest had yet to be exploited; a timber panic was not yet in the offing.[17]

Yet Sargent's evidence, when read against the historical record of forest devastation in other sections of the United States, suggested the grim outlines of the future, which is why worries about the extent of clearcutting found their parallel in uncertainties about how to restore cutover land. But the prospects for forestry in the United States looked bleak. The U.S. federal government lacked the kind of legal authority that enabled European governments to control private landowners; it was, moreover, required by federal law to give away the public domain and therefore sell significant amounts of valuable forested public lands on which it could wield power, and it lacked technical experts. Even though Congress had established the Division of Forestry in the Department of Agriculture in 1881, with Hough as its first head, the small agency had no woods under its control—and would secure none until 1905. The new division's energies were also deflected by internecine struggles that embroiled Hough, his subordinates, and superiors and led, in the summer of 1882, to Hough's demotion to the status of an "agent" of the Department of Agriculture. "Feel very low spirited," he scribbled in his diary, "and all my ambition is gone." His depression deepened when he and a colleague had to suspend publication of the

*American Journal of Forestry*, which they had launched only eleven months earlier. Doubtful that anyone read the voluminous reports he had written while in office and convinced that no real change had occurred in the nation's consciousness about the dangers timber devastation posed, he was left to compare unfavorably the differences between European forestry and American lumbering. Those differences once had inspired him and other reformers to enter into the political arena; now they were a painful reminder of how much remained to be done.[18]

Hough was a bit hard on himself. He and others had pioneered a new profession and created its first national voluntary associations, including the American Forestry Association and the American Forest Council, which would subsequently merge with the American Forestry Association. They had published the initial accountings of the status of the American forests, tried to raise the nation's consciousness about the evils of uncontrolled lumbering, and even gained some governmental recognition of their concerns through the creation of the forestry division. Without these contributions—however tentative and incomplete—the labor of those who followed would have been that much more difficult.

Among those indebted to this pathbreaking work was F. P. Baker, who, two years after Fernow traveled to the United States, set sail for Europe as one of the U.S. commissioners to the 1878 Paris Universal Exposition. In the opening to his report on the exhibition of European forestry, he doffed his cap to those on whose scholarship his scant knowledge of the subject depended; he readily acknowledged that his account was "hampered by the reflection that already the whole subject has been ably treated by writers who have brought to bear upon it the resources of immense observation and profound scholarship." Mindful of the attainments of Marsh and Hough, Baker preemptively limited the significance of his observations to an updating of "the history of forestry in Europe to a later period" and the addressing of "hitherto unreported progress." More important, he believed, would be the opportunity once more to "impress upon the American people of the United States the vital importance of the subject of forestry." If they were at all engaged by his account, he would "have accomplished all that can be reasonably expected" of him.[19]

The 1878 Paris fair was itself impressive: like other international expositions of the late nineteenth century, it offered participating nations the opportunity to flash their commercial wares, tout their industrial might, and demonstrate how they exemplified the virtues of rationality, progress,

**F. P. Baker visited the French pavilion of forestry at the 1878 Paris Universal Exposition. The structure was built entirely of "200 varieties" of wood grown in France.**

and civilization—values that, as the event's title asserted, were presumed "universal." Emblematic of these cultural displays was the magnificent French pavilion on forestry, known as The Chalet. Built "entirely of woods grown in France, at least 200 varieties being used in its construction," the striking edifice housed geological and entomological collections maintained at the French national school of forestry, as well as "maps, plans, photographs, and models representing the processes of reforesting mountains and of retaining the shifting surface of sand hills." Most stunning was a Disney-like model of mountain forestry that was set within The Chalet: trudging up a "zigzag track" carved into a fabricated hillside were workers "dressed in the peculiar costume of the country"; on their backs were heavy sleds. When they reached the peak, they loaded logs on the sleds and sent them down a "timber slide," which curved past charming representations of "the natural features of mountain scenery, the yawning ravines and plunging water-courses…." A stunning reflection of the French attention to detail, this display, and others that graphically demonstrated the national investment in reforestation and afforestation, convinced Baker of the great benefits to be derived from "French skill and industry," through which "every foot of earth" was carefully "preserved and patiently and laboriously cultivated." Improvident Americans had much to learn from their thrifty European compatriots.[20]

Whether Americans could replicate the successes of French conservation was another matter, for governing land management in France was a set of strict laws that granted governmental jurisdiction over the forests of the public domain, as well as communal woods, and private woodlots; its administrative structures, rigorous training for foresters in the national forest schools, and stiff penalties associated with timber poaching or careless handling of the forest amazed the American commissioner. The existence of a rational landscape and disciplined people, Baker quoted a French writer approvingly, enabled the "forest corps" to take "its place beside the great public services" whose futures would be determined "by the scientific skill and industry which characterizes our age."[21]

But to Baker's disappointment, neither law nor custom controlled industrialization's excesses in the United States. Although he did not propose that his native land adopt uncritically European legal models and technical training, he was clearly fascinated by how the French, Germans, British, Scandinavians, and Swiss were able to regulate the exploitation of natural resources so as to build up their national treasuries even while protecting the land. By contrast, Americans were "famous destroyers of the forest," devastating "[t]housands of acres of noble forest trees…merely to rid the earth of them. The Western pioneer," Commissioner Baker concluded, "has passed his life in toilsome labor of chopping and burning trees which his descendants would gladly replace."[22]

Absent a strong central state, and without cultural support for conservation, what "can we do to preserve and restore our forests, to repair the waste of the past, and provide for the needs of the future?" Baker's answer was couched in language that revealed the limitations he believed would prevent European forestry—for all its strengths and prospects—from being easily transplanted to American soil. Acknowledging that in the United States of late "a growing sentiment has sprung up in favor of the preservation and cultivation of trees both for ornament and use," he was not yet convinced the public would support the level of governmental "interference" that gave European foresters such unrestrained authority. He doubted, moreover, that legislative mandates were "efficacious" in any event: "few statutes have been more persistently violated," Baker observed dryly, than the already-extant ones prohibiting "cutting timber on government land."[23]

It is odd therefore that he put his faith in the U.S. Timber Culture Act of 1873, one of the most persistently violated of federal laws. Designed to encourage prairie homesteaders to plant trees on 40 acres of a 160-acre tract to meet residency requirements and facilitate subsequent ownership, it had no appreciable effect—except, that is, in wooded areas, where the act became a vehicle for fraudulent claims that enabled corporations to clearcut timber without charge. Still, Baker was convinced that what salvaged this act was its political acceptability: it did not promote forestry through "repressive" means, but by holding out "substantial inducements for the cultivation of trees, [it] becomes the patron and encourager of forestry, and thus fosters a popular sentiment in favor of tree-growing. Democracy could be pushed only so far."[24]

For Baker, push came to shove in 1884. That year, in a speech to the American Forestry Congress, he advocated the withdrawal from sale or entry of federally owned forests draped on either side of the Rocky Mountains so as to protect the headwaters of the Platte, Rio Grande, and Arkansas rivers. Drawing on his earlier observations of European forestry, he now proposed a multifaceted plan to preserve and manage this vast landscape, including the development of schools of forestry and regional forestry experiment stations, and the funding of federal surveys to establish the inventory and value of these timbered lands. He also called for protection against fire and illegal cutting, and the end to low-cost sales of wood harvested on public lands. "Government timber," he advised, "should nowhere be sold at $1.25 an acre. If sold at all a price should be fixed upon it somewhere near its value." Like Hough before him, Baker had reached a point where it was no longer acceptable simply to work within current political constraints. Because of the relentless destruction of the nation's woods, the time had come to mount a more sustained challenge to the status quo.[25]

That challenge would be taken up, and with considerable success, by a new cohort, many of whom were professional foresters who either had emigrated from Europe (Bernhard Fernow and Carl Schenck) or had been trained there (Gifford Pinchot and Henry S. Graves). These men may not have agreed on all points about the course of an American forestry, but as educators, reformers, and activists, they taught a wider public about the value of forestry to the commonweal. In their capacity as civil servants,

Fernow, and later Pinchot and Graves, also built the bureaucratic apparatus and secured the requisite political authority to establish a national forest policy. This crucial organizational work was reinforced through their contributions to myriad professional organizations; it is significant that one of these, the Society of American Foresters (1900), supplanted the older American Forestry Association as the leading voice on forestry affairs.

That said, their achievements were predicated on the previous generation's intellectual commitments and political activism. It was Marsh, Hough, and Baker, among others, who were the first to actively engage in the exciting trans-Atlantic exchange of ideas, who sought out and popularized European forestry; it was they who discovered just how complicated it would be to introduce its principles to an industrializing America. In sparking a civic debate over the future of the nation's forests, they proved instrumental to the coming struggle to restructure the nation's government so as to develop new controls over public lands. These early "lovers of the forest," Pinchot later confirmed, "deserve far more credit than they ever got for their public-spirited efforts to save a great natural resource."[26]

❦

# CHAPTER TWO

# *Why* Garden and Forest *Mattered*

Stephanie B. Sutton, biographer of Charles S. Sargent, made no extravagant claims for *Garden and Forest*. Sargent, a Harvard professor and director of the university's Arnold Arboretum, published the influential magazine between 1888 and 1897. Acknowledging that *Garden and Forest* was a "first-rate publication" deftly edited by the talented journalist William A. Stiles, and conceding that it was "an immediate success," Sutton drew a line between the praise it received and its lack of "popularity." Although the plaudits were "gratifying," a larger circulation would have eased the journal's chronic state of financial crisis. This tension shaped Sutton's assessment of *Garden and Forest*'s importance. Its readers included "people who worked with plants—foresters, nurserymen, botanists, landscape designers, and others whose opinions mattered." The arenas in which their viewpoints surfaced were scientific and political. By its coverage of things botanical, the "magazine did its small part to raise the horticulturalist from amateur to professional standing." Its words, "when quoted in a politician's speech or in the column of a popular newspaper,…carried authority."[27]

By so arguing, Sutton suggested that the magazine had an indirect influence, and that the enthusiasm its launch generated must be framed by the knowledge of its early demise. She reinforced this point when she cited the tribute *Garden and Forest* received from William Carruthers of the British Museum: "We are all pleased very much with your journal," he wrote to Sargent in October 1888, "and I hope you long live & maintain it at its present high level." Such was not to be.[28]

Yet by structuring the history of *Garden and Forest* as a narration of opportunities lost, Sutton failed to recognize the magazine's singular significance: it was the most vital late-nineteenth-century forum in which the debate over the place of science in public affairs was aired and the social function of the scientist was advanced. Those who contributed its articles on horticulture or landscape design or forestry were not just

As the magazine's editor, William Stiles helped establish *Garden and Forest* as a leading voice of the emerging forestry movement.

conveying the latest discoveries in their respective fields, but were writing themselves into a larger cultural discussion about the role of experts in scientific research and the power of that expertise to shape public policy. Their insights would have profound national consequences, altering the conception of the federal government and shaping the contours of modern America.

Contributors to *Garden and Forest*, for example, boldly used the journal to assert their understanding of and support for a new agenda that would define the nation's natural resource politics. Nowhere was this more evident than in the proclamations about the emerging forestry movement, a movement with which publisher Sargent was intimately involved. His editorials calling for the preservation of the nation's forested lands, and editor Stiles's aggressive pursuit of those foresters (such as Bernhard Fernow and Gifford Pinchot) who shared these sympathies, established *Garden and Forest* as the voice of those demanding closer federal supervision of the public domain.

At the time, that demand was a tough sell, as the *Garden and Forest* staff and writers recognized. Americans had little sustained interest in conservative land management, and even less in the creation of a powerful bureaucracy designed to regulate resource exploitation—especially as long as timber was crucial to the industrial revolution. Softening the opposition to their arguments favoring the implementation of federal conservation regulations would mean changing public opinion, a task *Garden and Forest* took on with evident enthusiasm. In its ten volumes, the journal published an astonishing number of articles on forestry—more than 450 in all; together, these pieces proved essential to the education of a culture that hitherto had known so little about the profession, its scientific claims, and its social implications.

Europeans, by contrast, knew a great deal about the subject. They had long experimented with forest management, had developed professional schools to promote research and educate succeeding generations of forest officers, and had pursued this work within political cultures— either monarchical or republican—that granted governments considerable power over public space and private property. Since the late sixteenth century, James C. Scott argues, some Europeans had been learning to see like a state.[29]

Americans did not begin to adapt that capacity and language until the mid-nineteenth century, and it is telling that among the first to embrace this European insight were those most anxious about the rapid disappearance of New World forests. Encapsulating their fears was George Perkins Marsh's pathbreaking *Man and Nature: The Earth as Modified by Human Action* (1864), written out of his years spent studying despoiled European environments. Like Marsh—who was an important influence on *Garden and Forest*—Sargent also focused on European models. In looking eastward, its editors, writers, and readers joined with other reformers in participating in a vibrant cross-Atlantic exchange of ideas. These Americans, whether concerned about urban social services or rural communities or devastated wildlands, readily absorbed European definitions of the issues and prescriptions for how to resolve pressing social ills through governmental intervention.[30]

Almost every number of *Garden and Forest* contained insights from abroad. One squib that editor Stiles inserted brought news of the formation of the Société des Sylviculteurs de France et des Colonies. Why publish such a small announcement that appeared to have little relevance to an American audience? Because the French group's stated ambition—"for the purpose of diffusing the knowledge of silviculture and increasing the popular interest in this art"—mirrored that of the magazine's patrons and subscribers.

Charles Sargent in the library at the Arnold Arboretum examining Quercus herbarium specimens. Sargent's *Garden and Forest* editorials helped influence national forest policy.

Stiles also made space for lengthy book notices touting the latest European advances in forestry science and devoted columns to enthusiastic first-hand accounts of tours of forests in Russia, Italy, England, Germany, and France. Consuming the most space, though, were learned assessments of the adaptability of the European experience to the New World. The authors of these articles themselves embodied the possibilities inherent in cross-cultural fertilization: Carl A. Schenck, who managed George W. Vanderbilt's Biltmore forests after Gifford Pinchot; his fellow countryman, Bernhard Fernow, who served as the third head of the U.S. Division of Forestry from 1886 to 1898; and American Pinchot, who trained at L'Ecole Nationale Forestière in Nancy, France. Each forester was sensitive to the difficulties in importing cultural institutions, but each knew that the American prospects for forest management initially depended on European ideas.[31]

Fernow and Pinchot in particular beat the drum for an enhanced governmental authority over public lands, and Sargent echoed their claims in his editorials calling for the creation of national forest reservations. Moreover, these men's educational work would in time nurture a growing cadre of like-minded professionals, an educational outcome the journal heartily supported. As such, they began the process whereby the champions of this transplanted discipline could claim a distinct niche within the American scientific community. "If I say that forestry has nothing whatsoever to do with the planting of road-side trees, that parks and gardens are foreign to its nature...that scenery is altogether outside its province," Pinchot asserted in *Garden and Forest* in 1895, "I am making a conservative statement with which every forester will agree." Its connections with "arboriculture and landscape art" derive from the fact that "it employs to a certain extent the same raw material...but applies it to a wholly different purpose." American foresters, like their European counterparts, were staking out their professional turf.[32]

This assertion of professional independence and scientific specialization, when linked to the slow but significant growth of public opinion in support of a more powerful federal form of forestry, was invaluable to the development of a progressive movement devoted to the conservation of natural resources and the expansion of the nation state. It is of lasting significance that this new ethos, which would dominate twentieth-century American politics, found early, sustained, and vivid expression in *Garden and Forest.*

❧

# CHAPTER THREE

# *Wooden Politics*

Federal forestry in late-nineteenth-century America was something of a joke. No one knew this better than Bernhard Fernow, whose first day on the job as head of the Division of Forestry in 1886 was rich in Dickensian humor. After climbing flight after flight of stairs in the even-then historic Agriculture building, he finally reached the tiny forestry office, tucked in the attic. If its size and locale had not convinced him of the disdain with which his profession was held within the federal bureaucracy, then the attributes of his chief assistant did. Nathaniel Egleston—"a reverend, white-haired gentleman"—had been the previous division head and had been demoted to make way for Fernow. This less-than-ideal work environment was made all the worse by Fernow's belief that Egleston was incompetent. His "knowledge of the whole subject was even less than that of his predecessor [Franklin Hough]," Fernow observed, and his administrative abilities were such that he "was at his wits end [about] what to do with the [division's] munificent appropriation of $8,000" per annum. As for Egleston's assistant, Fernow said only that he was a political appointee with no scientific knowledge of or interest in forestry. Yet Fernow's new subordinates knew the central fact of political sinecures: when he came upon them that first morning, they were "cozily, but by no means amicably, ensconced in a little garret room with two small oval windows, quarreling as to whom the credit for their performances really belonged."[33]

Fernow fully expected to infuse a more serious note into the division's proceedings. That, after all, is why he reveled in the telling of this anecdote. The laughter it was designed to provoke, the sympathy for Fernow it was to elicit, helped distinguish him from that comic pair. So did the anecdote's punchline. Fernow's antidote for his office's languid and slothful air was simple: he introduced a typewriter. This was an "innovation highly resented by the two," not only because it disturbed their quiet but also because it signaled a sharp shift in orientation: there was work to be done—work that would be done in a rational and efficient manner, work that would be regularized and codified in ways that only a typewriter could then produce. With Bernhard Fernow, modernity had arrived.[34]

So, too, had a certain prolixity. In 1898, for instance, Fernow proudly noted in his final (and thirteenth) report as chief of the Division of Forestry that his record of publication had far outstripped that of his predecessors combined; in the ten years that Hough and Egleston had been in office, they had managed to produce only four annual reports. The ever-exact Fernow knew that *annual* meant "each year" and had published the requisite number. He tabulated his publications in another way, noting that during his tenure more than six thousand pages of reports, bulletins, circulars, and other forms of "propaganda for [the] more rational treatment of our forest resources" had poured out of the division's single typewriter. He had filled, moreover, nearly twenty thousand pages of letter books, "largely containing specific advice given to correspondents." So that none would think that this productivity was a waste of energy and funds, Fernow calculated that his costs per page ($24, "hardly a fair charge for expert writing") were twenty percent less than those incurred "during the preceding period of nonprofessional writing." Efficiency, diligence, and economy were the hallmarks of his administration, or so this telling of the history of federal forestry would imply, a telling that left no doubt but that Fernow had exorcised the ghosts of Egleston and Hough.[35]

There was more to Fernow's governmental service than the written word. The significance of his contribution to the forestry movement unfolded most fully not in the number of pages he composed as America's first professional forester, but in the meanings embedded within those texts. It is only by analyzing these many publications that we can begin to evaluate the philosophical principles that formed the foundation of his work. This analysis is important, for the principles that shaped Fernow's perception of the proper role of government in developing, maintaining, and controlling the American landscape were the product of his training in the Prussian Forestry Department and of his later immersion in the intellectual currents sweeping America at the turn of the century.[36]

He had had some sense of this even in his youth. Born in 1851 in Inowrazlav, Prussia, a favored son of an elite landed family, Fernow began his apprenticeship in the Prussian Forestry Department and then received advanced training for two and a half years at the department's famed

Bernhard E. Fernow was the first professionally trained forester to hold the position of chief of the U.S. Division of Forestry. He put the agency on a permanent scientific footing during his tenure from 1886 to 1898.

academy at Muenden, where he studied under G. Heyer, and worked at several of the department's forests. His extensive academic study and practical experience were of minor value to his adopted country, which had little interest in forestry.[37]

Without a field in which to practice his calling, in the decade before Fernow became head of the forestry division in the Department of Agriculture, he held a variety of unrelated jobs, continued his studies of North American trees and the economic conditions of the lumber-based industries, published articles, improved his English to the point where he was no longer embarrassed to speak in public, and emerged as a driving force in the fledgling American Forest Congress, which had been established three years before Fernow arrived in America. He became such a presence within forestry circles that he was the obvious choice to succeed Nathaniel Egleston as head of the Division of Forestry. But so difficult had been his struggle to establish himself in this country that two years later, in 1888, Fernow's response to a Yale undergraduate's inquiry about America's future need for "educated foresters" was blunt: "Qui en sait?"

Well, Fernow did know, and in his reply, he urged the young Gifford Pinchot to study forestry with an eye to its usefulness "in other directions," including "landscape gardening, nursery business [or] botanist's work...."

Pinchot's prospects, the head of American governmental forestry assumed, were as dim as had been his own.[38]

Fernow's present situation left something to be desired. Although chief of the forestry division, his title masked the bureaucratic maze in which he labored. Public lands were under the purview of the General Land Office in the Department of Interior, which meant that the forester had no forests under his direct control. The General Land Office might ask his advice about how to manage the forests, but that assumed that management was desirable and possible. It was not. When Fernow assumed office in 1886, there were, for instance, no federal forest system and no public lands set aside for the practice of his profession. Regulations would emerge in time, and Fernow was active in their initiation. This was as true for the Federal Reserve Act of 1891, Section 24 of which enabled the president to "set apart and reserve, in any State or Territory having public land bearing forests…as public reservations," an addition to the bill for which Fernow took perhaps more credit than he deserved, as it was ensuing ancillary legislative initiatives that sought to protect these new forest reserves. By the early 1890s, approximately eighteen million acres had been set aside, a total President Cleveland doubled in 1897 when he proclaimed the Washington's Birthday Reserves. The concept of national forests, and by extension of forestry, was coming of age.[39]

Those dramatic changes notwithstanding, Fernow's work in Washington remained largely a matter of giving advice and serving as a conduit of information about forestry. Alterations in the numbers and sizes of forest reserves did little to change his earlier, gloomy assessment of the forestry division's activities: "under present conditions," he had written to a friend in 1887, "no practical work will be done and we might as well satisfy ourselves, that all we can do is talk."[40]

Yet for Fernow, talk was not cheap. The radical quality and political impact of his ideas emerge most clearly in his public discourse and private musings about how a system of forest management ought to develop and function in the United States. Never just an applied science, forestry was also an art whose success would require the reformulation of the philosophical basis of American political life.[41]

The first step—acquiring the plans for systematic forestry management—was simple enough. The United States need but look to Europe for its models. It is not surprising that Fernow, educated within the German forestry system, would believe that its methods were the most effective

**Chief Bernhard E. Fernow at the Division of Forestry exhibit at the Columbian Exposition in Chicago, 1893.**

and most culturally adaptive. They were, he observed, as applicable in British India as in Japan, two diverse Asian societies that had hired German foresters, adopted their silvicultural strategies, and then created forestry schools modeled after German educational programs to perpetuate this methodology. There was little reason why the United States could not follow suit, reaping the benefits of his homeland's technology and expertise.[42]

Fundamental to this transfer of knowledge was the adoption of a set of principles of forestry management that assumed a central place in what Fernow called his "propaganda work" on behalf of the federal government. In report after report, bulletin after bulletin, he argued that forestry was best understood as the Germans had defined it: "forest growth is to be treated as a crop to be reproduced as soon as harvested," and thus involved the idea of a "continuity of crops." In this, foresters were like farmers. Husbanding the "natural forces and conditions upon which the thrifty forest growth relies" was also part of the forester's charge, however, for it was no less critical that foresters produce "the largest amounts of material (or revenue) in the shortest time without impairing the condition and capacity for reproduction of the forest." That made foresters more like bankers than farmers, for proper forest management "involves the curtail-ment of present revenue for the sake of a continued greater revenue in the future"; this in turn required "continuity and stability to a greater extent than agriculture." Timbered land, then, was "permanently invested capi-tal, from which only the interest is used." Yet one could never cut into this capital: "the amount harvested or the revenue to be derived" should be "as nearly corresponding to the annual accretion" as possible. Good foresters balanced the books.[43]

To set up such accounts and establish the principles on the ground required intensive planning, as the Germans' experience revealed. During the early nineteenth century, Fernow reported, the various Germanic states had begun to invest heavily in topographical surveys of state-owned lands to record boundaries and typography, but also to determine the location of markets, and thus the cartographic relationship between forests and consumers. Onto these maps were then platted forest districts that established administrative and supervisory lines of authority running between the *Oberlandforstmeister*, or director of the agency, and the *Försters*, or district rangers.[44]

That organization extended to the very construct of the forest itself. A crucial task of the Försters was to conduct a survey of the district that would be carried out "to the utmost minutiae." Each district—especially those in relatively flat terrain—would be divided into "oblong compart-ments" of sixty to seventy-five acres, along each side of which, and at evenly spaced intervals, were cut a series of "openings or avenues" that ran north and south, east and west; each of these received a particular alphabetic designation depending on its orientation on the compass. At the intersections of these avenues, the Förster would place "a monument of wood or stone" that carried the identifying marks of compartment and avenues, "rendering it easy to find one's way or direct any laborer to any place in the forest." In Fernow's revealing commentary, this structure gave the German landscape the look "of an American city regularly divided into blocks"—the forest as gridiron.[45]

That emphasis on what Aldo Leopold later denounced as the German penchant for "slick and clean forests" appalled generations of American foresters; such detailed plans would not go into effect in the National Forest System. But for Fernow, and the tradition of which he was a part, there was a larger point to the regularizing of the land. Once this was accomplished, then human activities upon it could be controlled and rationalized. This involved establishing a set of legal regulations that determined rights and uses of the forests, drafting fire protection policies, and ascertaining what Fernow called the forest's "arithmetical basis"; among other things, this meant evaluating soil conditions and conducting precise tree measurements to create a database from which to assess rates of growth, timber yield capacity, and future productivity. These facts permitted rational management, whose definition accommodated neither wilderness nor irresponsible resource exploitation.[46]

That was the kind of management that Fernow expected to import to the United States, though he was shrewd enough to know that the German experience, born as it was of different historical context, social conditions, and political circumstances, was not an exact model for the American republic. "We in the United States are fortunate in that we can learn from the experience and profit from the assiduous work of these careful investigators," even if "we may never adopt [their] admirable administrative methods." But of necessity Americans would embrace the "technical measures" of German forestry, for these were based on "natural laws and proved by experience"—hundreds of years of experience, Fernow was quick to point out—and thus were essential to the attainment of "proper forest management."[47]

Yet even the technical measures would not be adopted readily without a wholesale change in Americans' conception of the powers and obligations of government—local, state, and national. With this, Fernow happily elbowed his way into the then-raging debate over the appropriate relationship between the individual citizen and governmental authority. He had no sympathy with what he saw as the contending parties in this issue. Neither the arguments of the "individualists" nor those of the socialists meshed well with the principles that guided forestry and shaped Fernow's political philosophy.

The individualists, those for whom a social Darwinian conception of society held great explanatory power, were misguided, Fernow declared. "It will be part of my theme," he wrote in "The Providential Functions of Government with Special Reference to Natural Resources," an essay that contains the most mature expression of his political ideology, "to point out the danger and impropriety of considering the social development of man as closely analogous to, nay, as of the same order as the biological development of plant and animal." That analogy, favored by Herbert Spencer and his American acolytes, perpetuated the notion that humanity had little control over its environment—that biology was destiny.[48]

Not so for Fernow. He deftly separated the idea of biological evolution and social development and argued that the latter, at least, was well within our control. There were, he said, "two qualities by which the human individual differs from the brute, the head and the heart, the intellect and the soul, the reason and the emotions...." And these "have had, and will in the future have still more influence upon the social development of the race." This must be so, he concluded, for "if we content ourselves to accept these [biological] forces as the only ones now at work in shaping *social* development, we shall fail in understanding, explaining or directing that development."[49]

No human society could evolve under these conditions, he affirmed. Progress could not "depend or...shape itself entirely under the working of the natural law of competition," a law the individualists championed. True, Spencer and others believed that individuals would "independently of society, develop the social instinct" and would do so "sooner and with less friction if let alone." But Fernow believed that a laissez-faire approach was by definition absurd: "It is not very clear why such a result should occur, how the free exercise of competition is to produce cooperation, which is its very antithesis."[50]

How then to secure the necessary social cooperation? Not through coercion, Fernow argued, taking a position that set him apart as well from those whom he labeled "rational socialists," those who asserted not the principles of laissez-faire but of *faire-marcher*. Their detailed prescriptions for social improvement, their propositions "to hasten the millennium," depended on "making cooperation compulsory and reason rule supreme." Those goals, however laudable in the abstract, could be achieved only at a great social cost: the socialist alternative would only suppress "the individual as in a colony of ants," Fernow declared, "each existing only as a part of the whole." The forester could hardly accept such social regression.[51]

Happily, there was a third path to social cohesion, that of the "true democrat, in whose creed society, the demos, stands recognized as the supreme ruler with the ideals of progressive civilization as the goal of associated effort." This figure was confronted with a tricky balancing act, negotiating between the needs of society and its individual members. As Fernow put it, the democrat must give "all liberty possible to individual activity that does not interfere with the good of society," a good that included "the moral and intellectual development and material comfort of all its members, present and future." To fulfill this required the acceptance of a new understanding of government, required a recognition that government was not an evil, not something separate from its people, "but as a good created by [the individual] for the attainment of his highest human ideals." This was a government to which Fernow could pledge allegiance. This, he confirmed, "is the creed to which I subscribe."[52]

His subscription makes sense in more than just political ways. In granting that this was the only form of governance that could secure not only "social existence, but social progress," in affirming that this meant that government had certain "providential functions," Fernow laid the groundwork for the development of a welfare state whose hand was visible in directing the present and future well-being of its citizens.

Overt guidance was especially necessary in terms of a society's natural resources, particularly its forests. Timberlands were quickly harvested, he noted, an exploitation that was legitimate under the reigning economic theories, but which also resulted in environmental disaster: lumber interests razed this well-wooded land, leaving behind a landscape of stumps. At the mercy of the "unrestricted activity of private individual interests," Fernow concluded, the forest "is quickly exhausted, its restoration is made difficult and sometimes impossible, its function as a material resource is destroyed." No progressive nation should accept such devastation, and its only protection lay in the "exercise of the providential functions of the state to counteract the destructive tendencies of private exploitation."[53]

In this sharply etched critique of American political culture, Bernhard Fernow was not alone. He was one of many government scientists—with C. Hart Merriam and Frank Lester Ward, among others—who argued for a dramatic rethinking of Americans' fervid commitment to unrestrained individualism. These new professionals were convinced that their academic training alone had fully prepared them to understand the necessity for social restraint as well as to direct its evolution. That is one reason why Fernow had been so dismissive of Nathaniel Egleston and Franklin Hough, his predecessors at the Division of Forestry. Why, too, he had applauded the government of New Jersey when, in 1894, it appropriated funds for a study of the conditions of the state's forests and then entrusted this task "not to a commission of ever so respectable, intelligent and patriotic citizens…but to an existing bureau of technically educated men, who were equipped to do this work thoroughly and authoritatively." For

Fernow and his peers, government should not be left to amateurs.[54]

Neither should it have been left to those he called the "cheap men"—to politicians. Unfortunately for Fernow, he spent ten years rubbing shoulders with those for whom compromise was an addiction, men whose social graces and political perspectives were equally coarse. They more than anything else had inspired his quest for a "providential government" and were a daily reminder of how wide was the gap between the ideal and real forms of governance.

His job was a constant source of frustration; he spoke of it as a "leaden anchor" that weighed him down. Each time the division's budget was slashed or other bureaucracies encroached on his already small terrain, he was pressed down further, and his "vigor and enthusiasm" were sapped anew. Things went from bad to worse when congressional leaders periodically pressured Fernow to halt his beloved studies of timber physics in favor of conducting constituent-pleasing research on weather modification, and worsened still when each new secretary of Agriculture—he suffered through four of them—proved no more supportive than the last.

Not surprisingly, Fernow constantly spoke of retirement from governmental service, and yet even when this came to pass in 1898, his interests were frustrated. He had hoped, even expected, that his able assistant, Charles Keffer, would replace him. Instead, Secretary James Wilson selected Gifford Pinchot, for whom Fernow held no great love. Moreover, in his cover letter that was appended to Fernow's last published report from the Division of Forestry, Wilson made it clear not only that Pinchot was working in "distinctly different channels" from his predecessor, but that these met with the secretary's "full approval." Events had come full circle: Fernow, who had dealt roughly with those whom he had succeeded, now knew something of their pain.[55]

More damaging was an episode that accelerated his final resignation. In 1896, at the urging of a National Academy of Sciences commission on national forests, President Cleveland set aside more than twenty million acres of national reserves on which forestry would in time be practiced. Fernow should have been pleased with the president's actions, given their implicit endorsement of his professional concerns and the bright prospects they offered for finally establishing large-scale federal management of forested lands; these Washington's Birthday Reserves might have marked the debut of Germanic forestry in the New World.

That was not how Fernow interpreted the presidential proclamation, countering that it was injudicious and perceiving it more repudiation than triumph. An explanation for his response lay in the politics surrounding the decision to establish the National Academy of Sciences commission in the first place. Particularly galling had been the fact that the request had originated within the American Forestry Association at its 1895 annual meeting in reaction to Gifford Pinchot's allegation that the association, and indirectly Fernow, had failed to protect America's forested domain. Affronted, Fernow fought to rebut Pinchot's "harangue," pointing out that congressional regulation was forthcoming and that the formation of the commission might imperil the legislative process. His rebuttal was ignored. The association's executive committee called for presidential action, and in February 1896 the secretary of Interior signed a letter requesting the National Academy to establish a commission to evaluate the condition of the public lands.[56]

Fernow was no more pleased with the commission's composition than he had been with its creation. Although its members included academic and governmental scientists, some were more professional than others; one whose academic credentials were smaller than his ambition was large was the ubiquitous Gifford Pinchot, chosen as the commission's secretary. Then there was the chairman, Charles Sprague Sargent. He, John Muir (an unofficial member of the commission), and others demanded that the U.S. military protect the forest reservations and that the forests therein be forever preserved—twin blows to the very concept of civilian foresters and public forestry. Finally, there was an even more personal snub: Fernow was not selected to be a member of the commission. As he confided to one correspondent, "I have neither been consulted nor in any way asked to contribute my share, nor recognized in my existence as the representative of the Government of this question." Having chosen not to consult its resident expert, Fernow believed, the government would gain little from the commission's tour of the national reservations during the summer of 1896.[57] He could not have been more wrong.

President Cleveland's startling proposal to add twenty-one million acres to the forest reserves had dramatically upstaged Fernow's decade-long and painstaking labor to establish a national forest system by working within the labyrinth of congressional politics. Fernow also believed that by infuriating western representatives in Congress, the Cleveland reserves threatened what success he had been able to achieve. About this, he was right: fearful that the reserves would be closed to development, congressional leaders

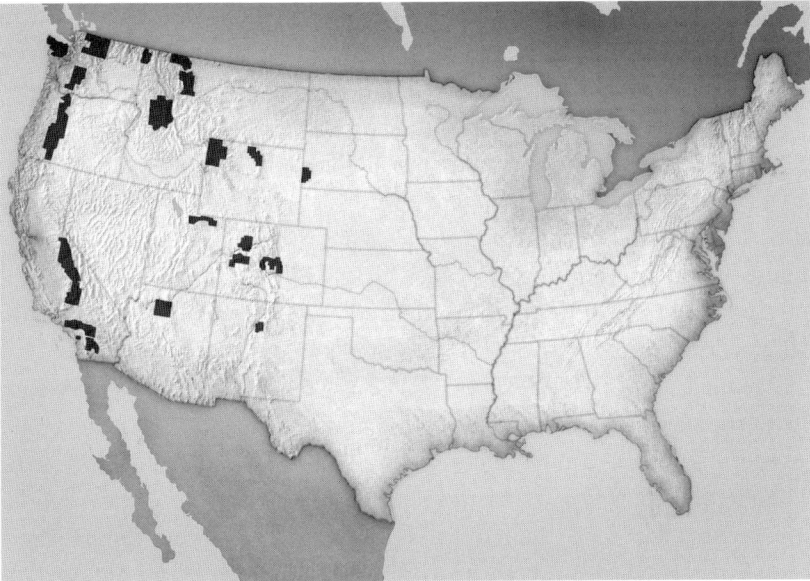

JUDY DERSH AND RUTH WILLIAMS, USDA FOREST SERVICE

**The federal forest reserve system at the time of Bernhard Fernow's resignation in 1898 consisted of more than thirty-four million acres of forest and grazing lands.**

moved swiftly to rescind the president's action. Ultimately the reserves were saved, and Fernow figured prominently in their salvation, but the whole affair deepened his disenchantment. He felt unappreciated within the federal bureaucracy and challenged, if not outmaneuvered, within the forestry movement that he had done so much to establish. He also seriously doubted that a full-fledged national forest policy, complete with a centralized system of forest management, could ever develop, given the fractured character of American politics, a doubt he readily shared with his successor, Gifford Pinchot. The government had not proved to be providential after all.[58]

Neither would Cornell University. In the summer of 1898, Fernow resigned as division chief to become the first head of Cornell's new state-funded school of forestry, believing that his vision for the profession could best take root in the academic environment. There, alas, Fernow encountered many of the same problems that had dogged his efforts in governmental service. Although the curriculum that he devised drew heavily on "the most advanced German ideas in forestry education," as did his working plan for the school's thirty-thousand-acre demonstration forest, his budget was never large enough to sustain his ambitions. The program was also understaffed. It did not help that the Pinchot-led Bureau of Forestry lured away Cornell faculty and snapped up most of its graduates, depleting the numbers who could work for Fernow. The chief blow, however, came in the guise of a crippling lawsuit filed against the school when it failed to produce the amount of lumber it had contracted to cut. In its wake, the state legislature withdrew the school's operating funds, thus killing the Cornell program in 1903, less than five years after its commencement. This effectively ended Fernow's career in the United States, too, for in 1907 he became the dean of a new forestry school at the University of Toronto.[59]

Cornell's collapse aside, Fernow could take comfort, ironically enough, from the fact that many of the ideals he had hoped to institute at the governmental and academic levels were taking hold. In 1905, President Theodore Roosevelt and Congress shifted responsibility for the national forests from the Department of Interior to the Department of Agriculture, established the Forest Service and its regulatory powers, and pumped up its budget. Fernow could take partial credit for all this, for although he had had strong doubts that this level of governmental management of forests would ever come to pass, he had been one of the first to articulate its necessity. Advancing his profession's status was his most important legacy: New World forestry was no longer a laughing matter.

CHAPTER FOUR

❧

# What Really Happened
# in the Rainier Grand Hotel?

One of the enduring narrative tensions that animates much of American environmental historiography is the conflict between those who argue for the preservation of natural spaces and those who call for their conservation and use. Made to stand as champions of these two apparently irreconcilable ideological positions are John Muir, a founder and the first president of the Sierra Club, and Gifford Pinchot, the first chief of the Forest Service. They had met in June 1893 in the New York City home of James and Mary Eno Pinchot, when Muir was fifty-five and Pinchot was thirty-one, and thereafter had maintained an engaged correspondence. When possible, they hiked and camped in some of the West's most spectacular landscapes. Their amiable relationship allegedly soured in the late 1890s because of sharpening differences in their perspectives on the human place in nature; these differences came to outweigh the benefits they had derived from their excursions into the wild.

No moment seems to have captured this rupture more vividly than a reportedly heated exchange between the two men in the lobby of Seattle's Rainier Grand Hotel, sometime in early September 1897. Please note the qualifiers in the preceding sentence: although many historians have recounted the incident in some detail and believe it signaled an irreparable breach in the men's personal relationship, and consequently in the two wings of the conservation movement with which they are so strongly identified, there is no incontrovertible evidence that it ever happened. I did not know that that was the case when I began work on a biography of Pinchot, in the pages of which—I must confess—I had very much anticipated narrating this tale because of its inherent drama and symbolic importance. It was only after I had read through the various secondary accounts that I noticed some small discrepancies that then forced me to sift through the

relevant primary sources, which do not completely confirm the substance of the story as it has been told.

That does not mean Muir and Pinchot did not have a confrontation at that time, just that the surviving records do not fully substantiate such a claim. Lack of certainty often drives my students to distraction: they want to know what happened *exactly*, and I can only shrug, making a physical gesture of some heuristic value. Of greater import, I hope, is the experience my students have in reading through the documents below, and wrestling with their many meanings and lacunae. I use this episode, for instance, to open an undergraduate seminar on environmental history, and on its first day we read aloud three versions of the story, initially narrated in Linnie Marsh Wolfe's Pulitzer Prize-winning biography *Son of the Wilderness* (1945). The students are encouraged to circle those names, arguments, and ideas with which they are unfamiliar, and to note alterations in the story as

The friendship between John Muir (opposite page) and Gifford Pinchot soured over conservation issues. When and how has been the subject of debate among historians for decades.

FOREST HISTORY SOCIETY COLLECTION

subsequent historians recast Wolfe's original construction of the altercation in the Rainier Grand lobby. We then discuss what they do not know or understand, work our way through the many discrepancies embedded within these texts, and analyze the literary strategies each author employs to make her or his case.

### First Narration:
### Linnie Marsh Wolfe, *Son of the Wilderness*[60]

*Stopping in a Seattle hotel on the way home, Muir read in a morning paper that Gifford Pinchot, then in the city, had given out a statement that sheep grazing in the reserves did little if any harm. Now, it happened that the most potent politicians of the state were either sheep or cattle men, or closely allied with live-stock corporations. So along came Mr. Pinchot with his little appeasement policy! This struck Muir like a* blow in the face. Not only did it shake his faith in a man he had trust-ed, but to his mind it augured ill for conservation, since Pinchot, recently appointed by Cornelius Bliss* as Special Forestry Agent, was here to make a survey of the commercial resources of the suspended Cleveland reserves.[†][61]

During the morning Muir saw Pinchot in the hotel lobby, stand-ing a bit apart from a group of men who—unbeknown to Muir—were newspaper reporters. According to William E. Colby,[‡] to whom he relat-ed the incident, Muir went up to Pinchot and, thrusting the printed page before him, demanded: "Are you correctly quoted here?" Pinchot, caught between two fires, had to admit that he was. "Then," said Muir, his eyes flashing blue flames, "if that is the case, I don't want anything more to do with you. When we were in the Cascades last summer, you yourself stated that the sheep did a great deal of harm."[§]

Thus the rift opened that swiftly widened between the two schools of conservationists—the strictly utilitarian, commercial group who followed Pinchot, and the aesthetic-utilitarian group who followed Muir—a rift that was to manifest itself deplorably in long years of antagonism between two Government bureaus.[62]

### Second Narration:
### Lawrence Rakestraw, "Sheepgrazing in the Cascades"[63]

The grazing controversy in the Northwest had a further signifi-cance in driving a wedge between the utilitarian and the recreational groups interested in conservation. Previous to this time, the strongest supporters for the reserve policies had been the recreational group. As a member of the Academy group, Pinchot had agreed with their views as to sheep grazing. By 1899, however, influenced probably by the Coville report, he changed his mind on the matter.** Muir learned of

---

* Cornelius N. Bliss was secretary of the Interior.
† On Washington's Birthday, February 22, 1897, President Grover Cleveland, just ten days before he left office, created thirteen new forest reserves, totaling more than twenty-one million acres; this act, which more than doubled the extent of the nation's forest reserves, set off an explosive political crisis.
‡ Colby was a close friend of Muir's and secretary of the Sierra Club; he is Marsh's sole source for the meeting, and his conversation about it with her makes her narration a third-hand account.
§ Muir and Pinchot had been together on the 1896 National Forest Commission tour of western states during the summer to determine how or whether the federal government should regulate the use of public lands and their natural resources.

*this about-face in Seattle on his return from the Harriman expedition,*†† *and was infuriated. His biographer has written on the episode: "Thus the rift opened that swiftly widened between the two schools of conservationists—the strictly utilitarian, commercial group who followed Pinchot and the aesthetic-utilitarian group who followed Muir—a rift that was to manifest itself deplorably in long years of antagonism between two government bureaus."*

### Third Narration:
### Michael Smith, *Pacific Visions*[64]

*That summer, Muir and Sargent*‡‡[65] *traveled together in Alaska, where Muir had ample time to hear Sargent's now substantial list of grievances toward Pinchot. Returning through Seattle in August, Muir noticed a local newspaper story reporting that visiting forester Gifford Pinchot had declared that sheep grazing in the national forest reserves would cause little, if any, damage. Pinchot, who happened to be staying in the same hotel as Muir, soon found an irate John Muir at his door demanding to know whether the article was correct and reminding Pinchot that during the previous summer he had voiced his agreement with Muir's strong opposition to grazing in the forest reserves.*

### The Story: Primary Documentation

An intense debate over those particular—even peculiar—iterations of the story usually ensues,[66] leaving the students not in a happy mood. "What's the point of writing about the past," a disgruntled junior once asked, "if you can't get it right?" Asking what would make it "right" is the only way I know how to use such questions to push the discussion

---

** Rakestraw makes two important alterations in Marsh's account: a date (however inaccurate), and a text—the Coville Report; in citing the latter, which analyzed the impact of grazing on mountainous terrain, he made Pinchot's alleged change of heart dependent not on his desire to appease regional power brokers, but on contemporary scientific evidence.

†† Edward Henry Harriman, railroad magnate and urban land speculator, sponsored a scientific expedition to Alaska in the summer of 1899; Muir was among those who traveled north.

‡‡ Charles Sprague Sargent, head of the Arnold Arboretum of Harvard University, was also a member of the Harriman Expedition and had been head of the National Forest Commission; in that latter capacity he had battled with Pinchot over what the commission should recommend to Congress about the governance of the new forest reserves. Sargent wanted to forestall commercial use and have the U.S. Army protect these vital public lands. Pinchot believed this approach was political suicide, arguing that no western congressional delegation would support such exclusion, so he pushed instead for the establishment of a professionally trained, civil Forest Service to protect and regulate the use of these valuable resources.

**John Muir and Gifford Pinchot disagreed over allowing sheep on the forest reserves. Although sheep presented one danger to the health of western forests, sheepherders posed another danger by setting fire to the forest to clear ground for driving. A forest ranger on the Sierra National Forest in 1900 photographed this burned Jeffrey pine as evidence of the harm unregulated grazing could do to forests.**

toward the primary sources that might confirm or discount some elements of this story. Students then begin to suggest the kind of evidence —newspaper records, diaries, correspondence—that might help determine the facts of the case. What follows are excerpts from the primary sources, and some of the questions that they help answer.

- When did Muir and Pinchot meet in the lobby of the Rainier Grand Hotel?

### 1. Gifford Pinchot's Diary:
### Seattle, Sunday, 5 September 1897

*Rose late, met with Holmes,[§§] and we went together to Port Blakeley and saw the great saw mill. Lunch with him at the Rainier Grand and after met John Muir in the lobby. Spent the afternoon with these two. Much delighted to see Mr. Muir again. Dinner with them…Church in the eve with Holmes and then dictated interview for the* Post-Intelligencer—*gave it to Mr. John W. Pratt, editor of the same. Dict.[ation] to his stenog[rapher]. Then more later with Muir and Holmes.*

Pinchot relates an amicable encounter on the day *before* the newspaper story appeared in print. Note that he dictated the "interview"—a convenient way to get his words published without correction or counter.

- What did Pinchot say about sheep grazing in the local press that may have so angered John Muir?

### 2. "The Forest Reserves," *Seattle Post-Intelligencer*,
### September 6, 1897, p. 8

*Pasturage may also be permitted by the secretary under suitable rules and regulations…*

That is the only sentence (and a partial one at that) that refers to grazing in the two-column article; one wonders whether Pinchot's qualified advocacy of regulated grazing would have fired up Muir's flashing blue eyes.

- Did the two men meet again after the publication of the article, and if so when?

### 3. *Seattle Post-Intelligencer*: Tuesday, September 7, 1897, p. 2, 5

*John Muir, the eminent scientist…left home yesterday by the steamer* Puebla…*the* Puebla *sailed for San Francisco at 8:00 a.m.*[***]

---

[§§] J. A. Holmes, who worked for the U.S. Geological Survey, was an old friend and associate of Pinchot's.

[***] Followup questions: If Muir's ship sailed at eight a.m., when would he have boarded? If earlier than its departure time, as surely he must have done, when did the edition of the *Post-Intelligencer* that carried Pinchot's words hit the streets? Would it have reached the hotel in time for Muir to have read it, and to have confronted a presumably awake Pinchot about the significance of what he had said?

## 4. Gifford Pinchot's Diary:
## Seattle, Monday, 6 September 1897

*Errands etc. Met Bailey of Dr. Merriman's staff.*[†††] *Got outfit repaired in part. Took 4 p.m. G.U. train east…*

It's possible that they met again in the lobby (or at Pinchot's hotel room door, as in Smith's account) very early that morning, but if so, Pinchot considered errands more worthy of mention.

- If their relationship had been so damaged after their Seattle interchange, why did their correspondence continue, and why would Muir calmly discuss sheep grazing, as in the following letter responding to Pinchot's request for advice on expanding national forest reserves in California?

## 5. John Muir to Gifford Pinchot: December 16, 1897

*My dear Pinchot: I was very glad to get your cheery hopeful forest letter in which you say you are not only confident of maintaining the present reservations but full of hope that we shall be able to protect them in the future and increase their area. I must say however that I feel rather discouraged about Bliss and Herman[n]*[‡‡‡] *that they should have wholly ignored last Years Commission and all they did + recommended + that Herman[n] should have thrown open the Oregon Reserves to Sheepmen and sheep, + should now be asking an appropriation for the management of the forests by timber agents. What such management will be, a long costly destructive experience shows. Still will hope for the best, + do what we can to head off blundering plundering money making officials.*

*I am unable to describe any unreserved Government lands in Cal[ifornia]. They are so mixed and invaded with patented lands here and there the task would be a long one. The region about Lake Tahoe southward to the Stanislaus River should I think be reserved—Also the Shasta Region though sadly devastated. I suppose the stripped lands could easily be acquired by the Government.*

*With best wishes for the New Year to Your parents and yourself.*

---

[†††] C. Hart Merriman was head of the Biological Survey of the Department of Interior.
[‡‡‡] Binger Hermann was commissioner of the General Land Office of the Department of Interior.

*[P.S.] I shall be glad to hear how you succeed in your forest plans. I look with little [hope] with Bliss and Herman[n].*

In that warm and personal letter, written three months after the alleged confrontation, Muir makes it plain, at least in this context, that he distinguishes between Pinchot's actions, which he favors, and those of the federal land managers Bliss and Hermann, which he denounces.

**Last Laugh**

The goal of this exercise is not to resolve all questions about a single event in the lives of two men who were critical to the development of the early environmental movement in the United States. Yes, their friendship and alliance ultimately would collapse—but not entirely over sheep. Instead, they broke during the early-twentieth-century debate over the construction of a dam in the Hetch Hetchy Valley in Yosemite National Park; Muir fought against it, and Pinchot was one of its advocates, in a conflict that has been even more frequently recounted than that of the Seattle hotel episode. But when Wolfe—and those who have followed her historiographical lead—read these later tensions back to 1897, and then forward to the 1930s to explain the then-contemporary struggles within the environmental movement, she robbed each moment of its rich and particular historical context. But I am only too happy to forgive her, for she has given me a wonderful means by which to remind my students, as they launch into a semester-long encounter with a host of primary and secondary sources in environmental history, that this period is a highly contested arena, and that that should shape how they read it. From the start, they need to be on the alert, being as careful in their handling of primary documentation as they should be skeptical of the interpretative claims of historical scholarship. This does not always come easily, but there is a moment later in the semester when the class regularly manifests its skepticism most vocally. That is when we watch *The Wilderness Idea*, a 1989 documentary that probes the entangled lives of Muir and Pinchot.[67] As the film narrator begins to recap Linnie Marsh Wolfe's version of events in Seattle, and suddenly the screen is filled with a herd of bleating sheep running across a meadow, the students howl.

In 2002, after a book talk in Portland promoting my biography, *Gifford Pinchot and the Making of Modern Environmentalism*, a friend handed me a copy of the letter below. He knew of my continuing search for evidence

Although there is disagreement over when the initial rift between John Muir and Gifford Pinchot occurred, the final break came during the debate over whether to construct a dam in Yosemite National Park's scenic Hetch Hetchy Valley, seen in these before and after photos.

pertaining to the incident at the Rainier Grand Hotel and chuckled with delight that he had located a document that raised anew the possibility that the two men had in fact discussed Pinchot's interview in the *Seattle Post-Intelligencer*. As such, this 1901 letter reminds us once more that historical scholarship, dependent as it is on the evidence at hand, will forever be revised as new documents from hitherto unknown sources come to light; and when they do, they will, likely as not, hoist the odd historian by his own petard.

### 6. John Muir to Charles Merriam: December 31, 1901

*My dear Merriam:*

*…As to the Oregon and Washington sheep pasturage. Pinchot said to me in Seattle when I called his attention to an interview in a morning newspaper that sheep did no harm in the Oregon and Washington forests.*[§§§] *This he printed in the principal newspapers soon after the Cleveland reserves were proclaimed, with a view to still the storm of protest, and thus he made the hearts of sheepmen glad. Coville too and indeed nearly the whole Bureau sing the same song, in some form, to avoid opposition and trouble from western Senators.—Talk of the harmlessness of sheepgrazing in forest reserves when "properly restricted", leasing forest reserves to sheepowners etc. under wise regulations—"the great importance of sheep industry." And of course you know that these forests are leased to sheepmen every year while almost all influential members of the Forestry Department bend to their wishes like saplings in the wind. But for "a' that" we'll win the fight in these states, as we have in California.*

---

[§§§] This still raises the question of at what time in the morning this interaction took place, or whether it took place the day before; note that Muir refers to Pinchot's views being published in the "principal newspapers," which suggests he may have learned of Pinchot's position not from the *Seattle Post-Intelligencer* but from another source, either before (or even perhaps after) their encounter in Seattle.

CHAPTER FIVE

❧

# Sawdust Memories

Henry David Thoreau once paused before a stump and meditated on the relationship between a woodchopper and a tree. From the stump alone, he wrote in "A Winter's Walk," "we may guess the sharpness of [his] axe, and from the slope of his stroke, on which side he stood, and whether he cut down the tree without going around it or changing hands." One could also tell, "from the flexure of the splinters…which way it fell," an insight that at once captured a precise moment in time and led Thoreau to extrapolate to all time: "This one chip contains inscribed on it the whole history of the woodchopper and the world."[68]

Thoreau was on to something, for the history of human interaction with forests can tell us a great deal about significant alterations in a society's politics, culture, and technology. The simple act of looking at a tree, and the language one uses to describe what one sees, depends on a broad range of available vocabularies. The technical and scientific grammar of biology, ecology, and forestry, which draw off one another and yet are different, would lead a viewer to describe a tree in varying ways, as would an economic, literary, or poetic frame of reference. The choices in this respect are endless and are made all the more complicated by the fact that such choices have changed over time. We do not perceive trees in quite the same way our ancestors did, just as contemporary loggers no longer swing an ax to fell a forest.

How then to decipher the inscription that Thoreau said lay within the chip? This question begs a related, and equally complex, query: What is the connection between forestry's past and its present? Any answer is made all the more complicated because the interaction between any two historical moments shifts over time and in part because establishing the connection depends on one's interpretative agenda. None of us remember the past—even a past we may have shared—in the same way. Yet how we remember forestry's history is critical, an importance particularly manifest when evaluating the relationship between Gifford Pinchot, arguably one of the nation's greatest woodsmen, and the profession he helped create.

To begin, there is the matter of just how good a forester he was. He called himself America's first scientifically trained forester, and said as well that in 1892 he instituted the nation's first forest management plan on George W. Vanderbilt's vast timberlands in Asheville, North Carolina; this plan, among other things, reportedly demonstrated the profitability of forestry to a skeptical American audience. Pinchot would make a virtue out of this litany of accomplishments when he penned his famous summary of his career: "I have been…a governor, every now and then, but I am a forester all the time—have been, and shall be, all my working life."[69]

But if he was, in his eyes at least, the creative force of American forestry, he oversaw a rocky creation. A close evaluation of his initial endeavors at Biltmore, for instance, suggests that his claims did not match reality. How could they? After all, his scientific training in forestry was sketchy at best, even by the standards of the late nineteenth century, a point made repeatedly by his mentors Lucien Boppe, with whom he studied for six months at L'Ecole Nationale Forestière at Nancy, France, in 1889–1890, and famed German forester Sir Dietrich Brandis, with whom he traveled throughout Germany and Switzerland. Their concerns were realized once Pinchot began work at Biltmore in early 1892. He was confused about where to begin, wrote lengthy appeals to American and European colleagues requesting detailed advice, and in a letter to Brandis confessed that "the time has come, as you foretold it would, when I begin to feel the scantiness of my preparation."[70]

Pinchot's forestry work also went slowly because the Vanderbilt forests had suffered from fire and overgrazing, which undercut his reclamation efforts. Moreover, the operation's labor force was untrained and its start-up costs were high, further frustrating Pinchot's initiatives; although he would say publicly that he had made forestry pay at Biltmore, in fact he operated at a loss.[71]

Those manifold problems would lead him privately to acknowledge having "done little in the work of my profession" during his first year there, and he was relieved to move on to New York to begin a career as a consulting forester; thereafter he monitored Biltmore's new forester, the much more thoroughly trained Carl Schenck. This shift in personnel relieved his European mentors, too: "the best thing that Pinchot has done," Brandis confided to forester William Schlich, "is that he secured Schenck for America." Pinchot's self-promotional conception of Biltmore

**Gifford Pinchot's effective self-promotion and family connections led to work for wealthy patrons in the Adirondack Mountains. Pinchot's consulting partner, Henry Graves, took this photo while they were working for William C. Whitney.**

as the cradle of American forestry was a piece of false labor.[72]

Pinchot would give birth to the real thing in succeeding years, but the reality he brought to life was not so much in the development of forestry in the woods as the establishment of its professional standing in society—a critical contribution to its future in a modernizing America. His timing was perfect, for it was in the late nineteenth century that other professions began to determine the requisite education and training, grant the appropriate degrees, and shape the behavior required of their future practitioners. The American Bar Association (1877), American Chemical Society (1875), American Historical Association (1884), American Economic Association (1888), and American Sociological Association (1905) established who could call himself a lawyer or chemist, a historian, economist, or sociologist—and, just as important, who could not. Pinchot would seek the same ends for the forestry profession while serving as head of the Division (later Bureau) of Forestry (1898–1905) and then as chief of the Forest Service (1905–1910). Between 1900 and 1905, he was a driving force behind the creation of the three major institutions without which no profession can exist—a professional society, graduate education, and a source of future work.

In 1900, at his grand home at 1650 Rhode Island Avenue in Washington, D.C., Pinchot called to order the first meeting of the Society of American Foresters, an organization that over time would bestow scientific legitimacy on, and structure the ongoing education within, the profession. Pinchot also had a hand in founding the *Journal of Forestry*, which would become the profession's leading journal, a publication that served to organize the value of the new knowledge it disseminated.

A profession requires professionals, naturally enough, and here again Pinchot was instrumental through the creation of the Yale School of Forestry in 1900. Together with the school's first dean, Henry S. Graves, whom Pinchot had tapped for the job, he contributed to the development of an appropriate curriculum in forestry education and occasionally taught within it. Because all professions must provide hands-on experience for their neophytes, Yale students spent their summers in Milford, Pennsylvania, training outdoors on Pinchot family forests, and indoors in classrooms the family constructed in the town. Having completed their training and received their degrees, young foresters would need gainful employment, and the bulk of them would be hired at the federal agency that Pinchot headed, the Forest Service. Given the significance of these labors, and the speed with which they unfolded—all within seven years— and given the agency's esprit de corps and sense of mission that Pinchot helped generate, it is hardly surprising that he thought he was the profession, and the profession was him.[73]

Pinchot's intense identification with forestry did not diminish during the more than three decades that passed between 1910, when President William H. Taft fired him for insubordination, and his death in 1946. Because he was so ineluctably bound up with the forestry profession, Pinchot felt compelled to plunge into the intense, internal debates that marked its development in the 1920s and 1930s, and simultaneously to confront serious external threats to its existence. This old forester refused to fade away.

Some of his successors at the Forest Service surely wished he had disappeared when they tangled with the former chief over changes in policy he did not believe were in the profession's best interests. Graves, whom Pinchot had handpicked to be his successor at the agency, was the first to feel the heat. Their relationship was strained when in the late 1910s Graves joined some of the younger leaders, such as future Chief William B. Greeley, in advocating closer relations with the lumber industry, an accommodation that Pinchot felt undercut the service's independence and integrity. Pinchot launched public rebukes of Greeley and others in a set of articles in the *Journal of Forestry*, the titles of which—"The Line is Drawn" (1919) and "Where We Stand" (1920)—cast the debate in black and white. They left little room for reconciliation, too, and the disputes intensified: throughout the 1920s, Pinchot continually attacked Greeley for mismanagement of the national forests and lobbied against Greeley-supported legislation, such as the Clarke-McNary Act, that he felt were too generous to the wood products industry and that violated what he perceived to be the public trust. Upholding that trust was foresters' first duty, he wrote, whether they worked in the federal, state, or private sector, for they were stewards of the land—a stewardship he felt was increasingly honored only in the breach.[74]

Stung by Pinchot's attacks, Greeley and others sought to isolate him within the profession, asserting that his charges were politically inspired. Their counterattack was effective, and by the late 1920s, no longer a member of the American Forestry Association, Pinchot also stopped attending meetings of the Society of American Foresters.

The New Deal gave Pinchot fresh currency. Casting about for a way to employ large numbers of the jobless in federal relief projects, in early 1933 President Franklin Roosevelt called on Pinchot, then governor of Pennsylvania, to submit a memorandum about how to use the national forests for unemployment relief. Drafted largely by foresters Raphael Zon and Robert Marshall, whose radical prescriptions for the restoration of both a battered landscape and a bewildered people dovetailed with Pinchot's gubernatorial initiatives, in which he had successfully linked conservation with job creation, the memorandum rejected past forest management policies. It argued that the most effective remedies lay in the immediate "large-scale public acquisition of private forest lands," tight public regulation of harvesting practices, and the hiring of sizable numbers of the rural and urban unemployed to work these newly acquired lands. Doing so would allow the federal government to assert greater control over the nation's natural resources, even as it restored the environment and the economy.[75]

Some of those ideas reappeared in legislation creating the Civilian Conservation Corps, and some found their way into the Forest Service

**Before the rift: In 1933, President Franklin Delano Roosevelt (seated third from right) visited the first Civilian Conservation Corps camp in Virginia with Secretary of the Interior Harold L. Ickes (seated third from left) and Secretary of Agriculture Henry A. Wallace (seated second from right).**

agenda, when Ferdinand Silcox became chief in the early 1930s. A liberal —but not, as one contemporary observed, "of the G.P. type"—Silcox was far enough to the political left to suit Pinchot; the two developed a close relationship. It was thus at Silcox's behest that the once-exiled chief returned to the fold to defend the fold, battling with the Interior Department and its pugnacious secretary, Harold Ickes, who schemed to transfer the Forest Service to the proposed Department of Conservation he was slated to head. Pinchot would combat this planned bureaucratic reshuffling with all the energy and aggressiveness he had displayed when confronting Greeley. Here, too, he had drawn the line.[76]

It was an old line, however, for there was nothing new in Ickes's demands, or in Pinchot's reactions to them. Dating from the original transfer of the forests to the Department of Agriculture in 1905, virtually every Interior secretary had sought their return; each time Pinchot had been a decided opponent. The most serious prelude to the Ickes-Pinchot wars had occurred in the early 1920s, when then-Interior Secretary Albert Fall campaigned vigorously to regain the national forests. Pinchot fought him in the press, challenged his actions on radio shows, and sent hundreds of letters to interest groups throughout the country and to every member of Congress; he even held a confidential interview with the secretary himself, the contents of which he then leaked to sympathetic newspaper editors. As effective as these assaults were, what really stopped Fall in his tracks was a scandal that derailed the entire Harding administration—the Teapot Dome affair. Pinchot had a hand in this, too: one of those credited with uncovering this sticky political mess was Harry Slattery, Pinchot's former secretary and close legal adviser. The two used these revelations of malfeasance to blunt Fall's efforts to capture the Forest Service—not the first time that charges of corruption in the Interior Department had worked to the agency's advantage.[77]

Ickes's quest began on higher, more defensible terrain. His desire to pull the Forest Service out of the Department of Agriculture, while a reflection of bureaucratic imperialism, was also related to his reasonable belief that a more unified, federal approach to conservation was essential. At the time, conservation programs and agencies were scattered throughout the federal bureaucracy, and their efforts were often at odds with one another, if their personnel even knew of the conflicting or overlapping agendas. Ickes convinced Roosevelt that a new Department of Conservation, into which would be folded most of the agencies of the current Interior Department, plus the Forest Service and others from Agriculture, would result in a more efficient and effective organization. His proposal meshed with the president's larger attempt to reorganize the whole federal government, resulting from work of the 1936 Brownlow Committee report, and Ickes's proposal therefore gained the president's sanction.[78]

To ensure the success of this reorganization, Roosevelt informed Secretary of Agriculture Henry Wallace that he could not work against the transfer. This order also muffled any voices of protest that might emanate from the Forest Service itself, which led Chief Silcox to turn to Pinchot for help. The older man—he was now seventy—was only too delighted to enter the fray; he had an ax to grind.

Once it was sharpened, he swung it freely, for he was as unconstrained by Roosevelt's agenda as he was by his once-close alliance with Ickes. The two had fought the good fight for the Progressive movement earlier in the century, but as the following years would demonstrate, there are no enemies like former friends. It was thus an intense personal disdain and political

President Franklin Roosevelt's attempt to let Secretary of Interior Harold Ickes "steal" the Forest Service from the Agriculture Department drew Gifford Pinchot into a very public battle with his old friend Ickes. This cartoon by Pinchot's friend, "Ding" Darling was published nationwide.

animus that drove Ickes and Pinchot during a long dispute that would not conclude until the mid-1940s. The battle was as fierce in public as it was in private, and they took turns pummeling each other through the press and over the airwaves, all the while assiduously lobbying members of Congress and even the president. In this Pinchot was not alone, and the struggle between the Interior secretary and his vociferous critics became so heated, one of Ickes's biographers has written, that it "proceeded along lines that were savage even by the standards of the often feral world of Washington bureaucracy." When it became clear that the opposition of the "Forest Lobby" threatened to disrupt the administration's legislative proposals, Roosevelt reneged on his pledge to support the transfer. Pinchot, who would die two years after Ickes's final failed attempt to recapture the Forest Service, left this life believing that he had preserved the most important piece of his self, of his past.[79]

Within the embattled agency, Pinchot's stout defense achieved legendary status. One of these tales, which I have heard over the years from several sources, revolves around the existence of a black box that allegedly resides in the office of the chief of the Forest Service. It is said to contain a set of files that detail, step by step, the actions the chief should take when threatened with the possibility of a transfer fight. The box's donor? Gifford Pinchot.

None of those who have related the story to me believe in it, but its apocryphal character does not diminish its significance. Granted, it gives the first chief a prescience that only immortals are said to have, but then such stories are designed to reassure the living; the dearly departed continue to look after their own.

The reverse is also true, for this reassuring tale offers a compelling example of the way the present can use the past to explain itself *to* itself. The anecdote is comforting, in this light, because it suggests that the contemporary Forest Service is not so different from its storied past, that it has remained more or less true to the perceived ideals and perspectives of its long-dead founder; where he stood is where his successors now stand.

Nothing is so simple. But the black box can serve as an odd mirror in which the present is able to catch at least a partial reflection of itself. It is this imaginative, conscious integration of memory into contemporary life and conversation that Thoreau had in mind when he stood before that decaying stump, wood chip in hand, and imagined that from this scene could he intuit the "whole history of the woodchopper and the world."[80]

**Divider photo:** *A Yale Forest School student practicing surveying in front of Grey Towers, the home of school founder Gifford Pinchot. The Pinchot family built a summer camp for the school on the 1,700-acre estate in 1900 and gave students unlimited access to the grounds for study purposes.* (Courtesy of Grey Towers NHS)

🌿

# CHAPTER SIX

# *Eminent Domain*[*]

*"The idea of the University of the South and the idea of its splendid domain are inseparable."*

—Arthur Ben Chitty[81]

President Theodore Roosevelt rarely minced words, and his vivid keynote address to the American Forest Congress in January 1905 proved no exception. Well aware that the conference had drawn a glittering assemblage of the nation's economic, social, and academic elite, and aware, too, that it had been staged in Washington, D.C., to pressure a recalcitrant U.S. Congress to pass a series of legislative initiatives advancing the administration's conservationist agenda, Roosevelt welcomed the nearly two thousand attendees and then gave them their marching orders. "For the first time the great business and forest interests of the nation have joined together," he declared, "to consider their individual and common interests in the forest." There was much for the two groups to discuss, too. Contemporary news accounts of widespread land fraud on public lands in the West dovetailed with staunch legislative resistance at the state and national levels to regulations that would control the sale, dispersal, and management of the public domain. Bucking this trend were those who, like Roosevelt, believed that conservation would produce better land management and more appropriate stewardship. "You all know...the individual whose idea of developing the country is to cut every stick of timber off of it and then leave a barren desert for the homemaker who comes in after him," the president declared. "I ask, with all the intensity that I am capable, that the men of the West remember the

[*] This essay was written in memory of the late Stephen Puckette, long-time member of the faculty of the University of the South, and in honor of his wife, my cousin, Upshur Puckette; their love of the Domain has been contagious, and it was in their rambling home on Morgan's Steep in 1998 that I began to think about the impact of the idea of forestry on Sewanee. I am grateful for their support, for the help and guidance of Karen Kuers and her generous colleagues in the Department of Forestry and Geology, and for the invaluable aid of archivist Anne Armour.

sharp distinction that I have just drawn between the man who skins the land and the man who develops the country. I am going to work with, and only with, the man who develops the country. I am against the land skinner every time."[82]

Roosevelt expected his large audience to share his antipathy and his outrage, as well as his conclusion that the real "prop of the country must be the businessman who intends so to run his business that it will be profitable to his children after him." Adopting such a multigenerational perspective was critical, for the nation was at a crossroads of its own devising: "If the present rate of forest destruction is allowed to continue, with nothing to offset it, a timber famine in the future is inevitable. Fire, wasteful and destructive forms of lumbering, and the legitimate use, taken together, are destroying our forest resources far more rapidly than they are being replaced."[83]

How to forestall this complex, dangerous situation? The "remedy is a simple one," the president assured his listeners. If the forest congress would adopt resolutions advocating more conservative use of natural resources, and if the U.S. Congress finally acquiesced to the administration's requests for the creation of a national forest service to coordinate federal conservation management, then contemporary Americans and their progeny would be well and truly served. "I wish to see all the forest work of the Government concentrated in the Department of Agriculture," Roosevelt told the congress. "It is folly to scatter such work, as I have said over and over again."[84]

The convention took the hint, passing eighteen resolutions, one of which supported the creation of a forest service in the Department of Agriculture. The U.S. Congress did its part, too: within a month the legislature had signed off on a bill, to which Roosevelt gladly attached his signature, which transferred the national forest reserves from the Department of the Interior to Agriculture and established an agency, to be known as the Forest Service, to manage these lands. In his charge to its new chief, forester Gifford Pinchot, Agriculture Secretary James Wilson reinforced Roosevelt's arguments about the significant purpose of the national forests: "the permanence of the resources of the reserves…is indispensable to continued prosperity, and the policy of this Department for their protection and use will be invariably guided by this fact, always bearing in mind that the conservative use of these resources in no way conflicts with their permanent value." And when a conflict arose, Wilson

Vice Chancellor Benjamin Lawton Wiggins strongly believed in conservation and federal cooperation when neither idea was popular in the South. His vision saved the college's badly cutover forest.

concluded, "the question will always be decided from the standpoint of the greatest good of the greatest number in the long run."[85]

Because of its role in the creation of the national forests and Forest Service, the 1905 American Forest Congress is much celebrated as a transformative moment in the history of conservation in the United States. Much less well known was its impact at the local level, especially in the South, a region that was largely ignored during the conference; President

Roosevelt's tough-love admonitions about the West's need to reform itself were of a piece with the conference's general focus on the management (and mismanagement) of that region's forested estate.

Yet at least one southern attendee took Roosevelt's provocative words to heart; he was Benjamin Lawton Wiggins, vice chancellor of the University of the South, in Sewanee, Tennessee. In the published version of his speech to the congress, Wiggins applauded the president's assertion that "the forest problem is in many ways the most vital internal problem in the United States," accepted that the sole remedy for the impending timber famine was "the introduction of practical forestry on a large scale," and as an educator, shared Roosevelt's conviction that only "men trained in the closet [schools] and also by actual field work under practical conditions" could avert the coming calamity of a deforested America. If national awareness of the "economic peril is coming to be realized everywhere," Wiggins confessed, that knowledge was considerably less well diffused throughout the South. Even so, there were some "far-seeing men [who] are now convinced that something must be done to prevent diminution of water supplies, the occurrence of disastrous floods, and the almost inevitable and speedy exhaustion of the timber supply." Wiggins counted himself among their number, and rightly so: since 1900, and at his insistence, agents of the federal Bureau of Forestry had been managing the university's more than six thousand acres of woodland, making it arguably the first academic environment so regulated. His embrace of the principles of forestry had had a major impact on this small mountainous community in southeastern Tennessee. When we explore the ramifications of his actions on the campus woodlands and its cultural identity, it becomes clear that at the University of the South, forestry was not just about trees; its greatest good may have been human.[86][†]

### A Promised Land

When Wiggins became vice chancellor of the University of the South in 1893, he entered a landscape—natural and culture—that was thick with memory. Sited atop a spur of the mid-Cumberland Plateau in Franklin County, with elevations ranging from eight hundred to nearly two thousand feet above sea level, the rugged terrain, and especially the caves in

---

[†] Since its inception, the University of the South has also been known as Sewanee, and I will use the terms interchangeably.

the escarpments that fall away from the plateau, had been semiperma-
nent homes to hunter-gatherers whose presence dates back to 8000 BCE.
By the 1820s, white land speculators had begun to lay claim to this high
ground; because the soils did not sustain extensive agriculture, settlers
grazed animals—cattle and hogs—as a food source. Timber harvesting
accelerated when the Sewanee Mining Company began its coal operations on the plateau in the early 1850s. Extracting this valuable resource
depended on the construction of a spur line down to Cowan, where it
linked with the Nashville and Chattanooga Railroad, a route that tied the
once-isolated mountain to the larger forces then shaping the antebellum
southern economy.[87]

The plateau was also inextricably tied up with the raging cultural war
between the South and the North that would later erupt into armed conflict. Five years before the Civil War, Episcopal Bishop Leonidas Polk of
Louisiana, after considerable thought about how best to train the rising
generation of southern gentry, decided that the region needed a college
to rival Princeton and Yale, with an academic environment better suited
to promote southern values, set "within the pale of the plantation south."
Those young men who went north for their studies, he advised his fellow southern bishops, traveled "beyond the reach of our supervision, or
parental influence, [and were] exposed to the rigors of an unfriendly climate, to say nothing of other influences not calculated, it is to be feared,
to promote their happiness or ours." As tempers flared in Congress and
mobs fought in city streets over the vexing issues of slavery and sectionalism, Polk was convinced that the establishment of a new university was
the only thing "that will save us as a church, and as a Southern Church
in particular." Because this new campus was to be the joint property of
all southern Episcopal dioceses, it had to be readily accessible. He assured
his correspondents that the best location in which to place the school—
to be known as the University of the South—was in Sewanee, near
Chattanooga and its multiple, intersecting railroads.[88]

Couched to appeal to southern nationalism, Polk's letter received warm
reviews. Most of its recipients supported his calls for a college—southern
in name and in deed—and after a series of conferences the bishops launched
an endowment campaign. Polk was convinced that it would prove a success, for regional religious sensibilities and academic aspirations would
impel philanthropists to support the cause. So, too, would the pressing
need to defend slavery: *The negro question will do the work*," he advised

Bishop Stephen Elliott of Georgia. "It is an agency of tremendous power, and in our circumstances needs to be delicately managed…. If we—churchmen—do not let it have its own way and operate through us, it will cast us aside and avail us of the agency of others." Coopting southern resentment of northern abolitionism would produce a strong, well-funded university that would rebut the northerner's sneer that "a slaveholding people cannot be a people of high moral and intellectual culture."[89]

Their elevated ambitions came into being in late 1857, when the bishops approved Sewanee as the preferred locale and accepted the generous offer of Samuel F. Tracy, president of the Sewanee Mining Company, to match an earlier pledge of five thousand acres of land from citizens of Franklin County; at ten thousand acres, the University of the South was one of the largest campuses in the United States. Tracy then sweetened his company's gift by promising one million board feet of timber, two thousand tons of coal a year for ten years, and free transport of twenty thousand tons of building material. The school's construction seemed assured, and the mountain, once a source of considerable mammon, would now be devoted to more sacred ends.[90]

Or it would have been had not the Civil War exploded. Although a cornerstone had been dedicated at elaborate ceremonies in October 1860 and a few homes constructed, the planning and development of the university were suspended because of the sectional crisis, or as the *Church Intelligencer* proclaimed, "until this unnatural and wicked invasion shall cease." Though far removed from the war's first engagement—the April 12, 1861, southern cannonade on Fort Sumter in the harbor of Charleston—its explosions immediately rocked Sewanee. That very evening, someone hurled firebombs into the Polks' hilltop home, where the bishop's wife and daughters were staying, and into the empty house of Bishop Elliott; both were completely gutted. Polk's family escaped thanks to the quick action of a domestic slave. "Was there ever in the all the world such a hellish proceeding," the bishop fumed from his residence in New Orleans. "I am satisfied that it was the work of an incendiary, and that it was prompted by the spirit of Black Republican hate."[91]

Polk was provoked in response to accept a commission as a brigadier general in the Confederate army; a West Point graduate who years earlier had resigned his commission to enter the church, the bishop now made haste to "buckle the sword over the gown." The "Fighting Bishop" returned to Sewanee once during the long campaign, staying at the site on July 3,

1863, as his troops staged rear-guard actions on the mountain to slow the Union army's advance on Chattanooga by chopping down trees to block passage along its narrow roadways. Later that day, Polk's forces retreated down the mountain, hooking up with the rail network into Chattanooga that had made Sewanee such an attractive site for the proposed university.[92] He was killed in action in May 1864.

The idea of the school did not die with Polk or the defeat of the Confederacy, however. Indeed, its prewar sectionalism was manifest in postwar campus life. The Reverend Charles T. Quintard, who had served with Polk in the Confederate army and would later become the second bishop of Tennessee, in the late 1860s sailed to England and there raised enough money to jump-start the new college. After hiring four faculty, he helped devise a curriculum that one historian has described as "a solid Anglo-Saxon mixture of British and southern elements, which looked to the past, to tradition, for inspiration." British in orientation, too, was the university's spatial design and architectural references: Gothic motifs dominated, "with some buildings being replicas of those at Oxford and Cambridge." Dress requirements reinforced the separatist mien: students wore Confederate-gray uniforms until 1892 and were drilled by former Confederate officers, and at least five members of the faculty were former generals in the Confederate army. Those who had been too young to fight, like Bishop Thomas Frank Gailor, a teacher and administrator at Sewanee, felt impelled to extend "the conflict's pervasive presence." His outlook was reinforced by the number of war widows who moved to the mountain to live in the surrounding community, a self-contained aerie in which to nurse the wounds of war and mourn the Lost Cause.[93]

William Alexander Percy, a 1904 graduate, captured the nostalgic haze that, like the region's legendary fog, enshrouded Sewanee, a community "presided over by widows and Confederate generals." Its altitude reinforced its retrospective gaze: "a long way away, even from Chattanooga," he wrote in his memoir, *Lanterns on the Levee*, the school is perched "on top of a bastion of mountains crenellated in blue coves. It is so beautiful that people who have once been there always, one way or another, come back. For such as can detect apple green in an evening sky, it is Arcadia…" Old times there could not be forgotten.[94]

## Shifting Ground

There was, however, nothing sentimental about the college's exploitation of its vast acreage, dubbed the Domain; Sewanee's economic actions were not nearly so conservative as its politics. Sandstone deposits were heavily quarried to build the main campus structures, and tons of rock was also sold to the Nashville, Chattanooga, and St. Louis Railroad to build its depots. Wood for fuel and building material was logged on site, coal deposits were mined, and livestock roamed freely. For a new school without a hefty endowment, the land's natural resources must have seemed bountiful, and so they also appeared to local residents, many of whom apparently used the Domain as a commons. Before the Civil War, the trustees had reacted to the repeated depredations by hiring a forest guard to patrol for trespassers and "prevent cutting down this valuable forest growth." After it, the thievery resumed: in 1880, Vice Chancellor Telfair Hodgson posted notices throughout the forest warning "against purchasing Wood from any person who does not exhibit Written Evidence of having purchased the same" from university-sanctioned loggers, "under pain of having the same confiscated."[95]

Neither guards nor signs seemed to have much effect. By the late nineteenth century, the woods had been high-graded, leaving behind only the poorest-quality timber; forest regeneration was compromised by extensive, unregulated grazing; and fires, whether from lightning strikes or human action, swept across the plateau, threatening the community's safety and adding to its environmental woes. Observed George R. Fairbanks, a historian of the university, "ignorant or willful wielders of the axe, disregarding all instructions or contracts…marred and destroyed large portions of the original forest growth." Whatever the impulse or source, the Domain had been so degraded that in 1896 a timber company proposed clearcutting the remaining forest cover for a paltry $2,000.[96‡]

In difficult financial straits, the university was tempted by the offer.

---

‡ Evidently Fairbanks was himself one of those who despoiled the Domain: in 1905, Vice Chancellor Wiggins reprimanded his colleague for hiring "two negro men [who] were caught cutting trees just beyond the Infirmary. They stated that they were acting under your instructions. As you know, we are under contract with the Forest Service to cut only such trees as are marked by representatives of the Service, and we have been strict in punishing all offenders. Ordinarily these two men would have been arrested; but as they would have placed the responsibility on you, I wished to refer the matter to you, and request that whenever you need poles that you will let me know, so that the trees may be properly marked and so that negroes may not be led to believe that they can make depredations at will." As a member of the faculty, Fairbanks had a moral responsibility, on a number of levels, to uphold university policy.

Built in 1878, St. Luke's Theological Hall as seen around 1889. The fences were erected to keep out the locals' livestock. The livestock ran throughout the Domain and caused extensive damage. About 1,000 acres of the campus were enclosed in a belated effort to protect the forest.

In the end, Vice Chancellor Wiggins persuaded the board to reject it. Jealous of the university's rights and prerogatives associated with Domain woodlands, within six months of his appointment to the university presidency, Wiggins was confronting trespassers on the land and in the courts. The university attorneys, Banks & Embrey of Winchester, cautioned the vice chancellor about being too aggressive: admitting that Wiggins was "pursuing the right course in getting after these offenders," T. A. Embrey advised, "We of course do not want any more lawsuits than are absolutely necessary to protect the timber on the Domain." In the case of one poacher, the lawyer suggested that the university settle the dispute on two conditions: that the miscreant pay damages and agree not to "trespass on your lands anymore."[97]

The strategy may have reduced the university's legal bills, but existing records indicate that the pilfering continued. Seeking a more effective strategy for controlling illegal resource use and promoting better management of the Domain, in 1896 Wiggins contacted consulting forester Gifford Pinchot. He may have done so because of Pinchot's prior experience managing the Biltmore forests, which like the Domain had been heavily logged, burned over, and badly grazed. That the forester had wrestled with many of the same land management issues that had so irked successive administrators of the Domain surely recommended him to Wiggins, who was searching for a more comprehensive and effective method of bringing order to an unruly terrain.[98]

The parallels may have also accounted for Pinchot's interest in the campus reforestation project. Yet the press of his consulting business interfered: "Mr. Gifford Pinchot, the well-known forester, continued to disappoint me in the long expected visit to Sewanee," a frustrated Wiggins advised the trustees in 1897, and "so we are still without expert advice as to the disposition of our timber and a proper care of our forests."[99]

Finally, in August 1898, Pinchot arrived on the mountain and was much struck by its manifold possibilities. "[H]e told me that the prospect for forestry was more encouraging than at any other place that he had seen except, perhaps, in the Adirondacks," Wiggins wrote to trustee Silas McBee, who had been instrumental in arranging the forester's much-anticipated and long-delayed visit. "We are all charmed with Mr. Pinchot, and I can't tell you how grateful I am to you for securing his interests and services."[100]

The wait had been worth it because Pinchot was no longer just an independent consulting forester. One month earlier he had been named head of the Division of Forestry in the Department of Agriculture, and his agency was in need of work. Because the nation's forest reserves were then under the purview of the Department of Interior, the small number of federal foresters in Agriculture had no public lands under their control. To practice their craft, gain publicity, and generate support for their work, in 1898 Pinchot issued a pamphlet, *Circular 21*, which encouraged private landowners to make use of the foresters' technical skills and managerial advice. "The program was immediately popular," historian Harold K. Steen has noted. Within a year, there were more than 123 "applications for assistance from 35 states, involving 1.5 million acres." Vice Chancellor Wiggins's request for the agency's aid could not have been better timed.[101]

### Working Plans

After his visit to the university, Pinchot hired Biltmore forester Carl Schenck as a special agent to prepare a full-fledged management plan for Sewanee. By July 1899, Schenck and five of his Biltmore students had prepared an ambitious working draft, containing a complete survey of the Domain. It segregated the more than six thousand acres into compartments that identified ecological niches and geographic zones; proposed the construction of an extensive network of roads for logging and fire protection; advocated the building of fire towers and a fence to enclose the woodlands; argued for the hiring of a full-time, on-site forester to manage the lands; and laid out a timber-harvesting cycle that Schenck estimated would net the university upward of $2,000 per annum. Together, these propositions were essential to the successful introduction of forestry on the plateau.[102]

Wiggins demurred, if only because he had to steward the university's too-thin budget. He made plain his disagreements to Pinchot when he forwarded Schenck's proposal to Washington. "I objected to certain items of expense which seemed unnecessary and would consume almost the entire profit from the sale of the timber," the vice chancellor reported to the trustees, indicating that Pinchot shared his negative reactions. The chief forester "recognized the fact that the wire fence and fire watchers were calculated to provoke trespass rather than prevent it, that the building of roads would not facilitate the transportation of logs to such an extent as to justify the large appropriation, and that the inspector who would be stationed here by the Bureau of Forestry, without expense to the University, would be sufficient." Rebuffed, Schenck was also replaced; Pinchot assigned one of his Washington assistants, John Foley, to take charge of the Sewanee operations.[103]

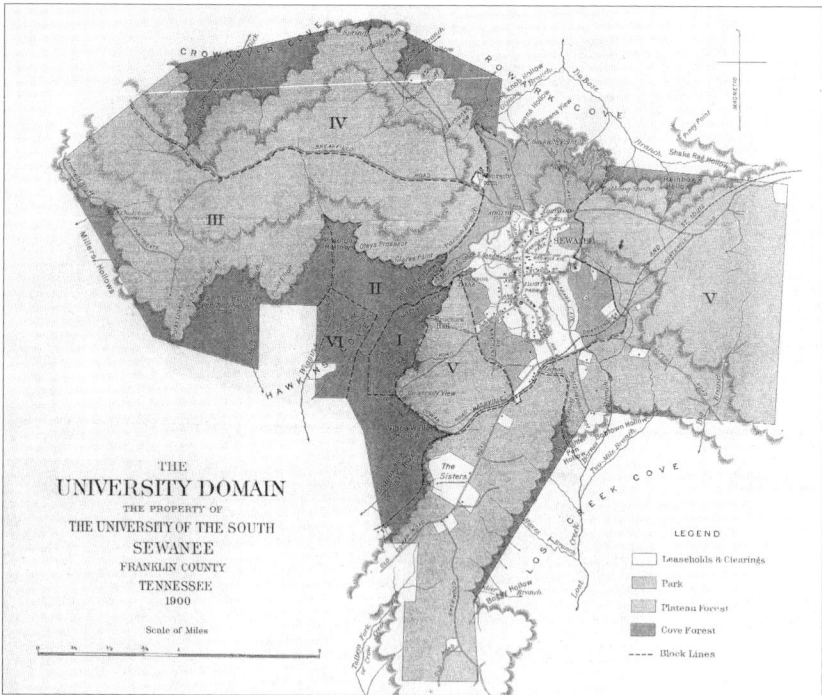

The U.S. Bureau of Forestry (later USDA Forest Service) mapped the University of the South's forest holdings in 1900 as part of its effort to provide a forest management plan. This map appeared in Bulletin 39, *Conservative Lumbering at Sewanee, Tennessee.*

Years later, Schenck would remember that his 1899 plan was "never executed, since the university was short of money. Forestry is no go with an owner short of money." But it was executed, if on a less expensive basis, following the same managerial means and achieving similarly profitable ends. Foley's report, *Conservative Lumbering at Sewanee, Tennessee* (1903), indicated that careful, regulated lumbering on the plateau and the coves (where most of the cutting had occurred) produced more than $2,000 in 1900–1901, slightly less the next year, and, he estimated, $1,500 in the coming years. "In a word, timber formerly valued at $3,000 will have been made to yield a profit of about $7,000." Forestry paid.[104]

Profitability was but one goal of the university's timber program. Another was a more sustained effort to manage the local population's activities. Foley gave voice to this when in his report he delineated which tree species could be harvested, when cutting could occur, and under what

The Domain as it looks today. The University of the South's campus and surrounding houses comprise about 2,500 acres of the total 11,000 acres of this mostly urban forest. Selective harvesting occurs on sixty acres a year in the predominantly oak/hickory forest.

conditions. It appeared as well in his recommendation that the university, "besides making every effort to create a sentiment against forest fires"—it had been a long-standing practice among plateau farmers to burn the woods to clear the land—"should be vigilant in extinguishing them." And it framed his conviction that the university's "lax forest management" in the past had encouraged "excessive abuse." With "no thought for its welfare," the school and Sewanee residents had pillaged a once-magnificent hardwood forest. Those attitudes and actions would change under the forester's tutelage; rational land management would make for a more rational people.[105]

### New South Prophet

Forestry would also change habits in another respect: its instruction would transform university curricula, alter the character of the student body, and reinvigorate college life. Or so Wiggins professed in his address to the 1905 American Forest Congress. In it, he detailed the rise of forestry education in America, paying special attention to the significant influence that the Yale School of Forestry had had on its university. That forestry science "is in active touch with the demands of practical life and the opportunities for employment," he argued, "gives the students of Yale an assurance that side by side with their training in general culture and public spirit, they are adapting themselves to speedy usefulness in the complex organization of modern commercial life." Recognizing that not all educators agreed with him, that many still resisted the introduction of scientific study in any form into college classrooms, Wiggins nevertheless believed that curricular reform was inevitable. "The world [is] moving on. New constituencies and new demands [are] arising, new problems [are] being projected on the economic and political horizons, new questions [are] pressing for answer."[106]

Forestry, as an academic discipline, was emblematic of the coming transformation in American higher education. Because a forester is "above all a man with practical problems to handle…he needs the democratizing influence of university life," Wiggins asserted. But for the same reason American campuses needed to offer forestry courses; their presence in course catalogues, and the students they would attract, "will cause our universities to come forth from their cloistered seclusion into a closer touch with the activities of life."[107]

Wiggins's assertions placed him in league with other New South reformers, those of the postwar generation who wanted to shake off the dead hand of the Civil War and the Lost Cause and believed that only a modern economy—efficient, rational, and planned—could revive the impoverished region. Collectively, they also sought, Paul Gaston has observed, a "lexicon [that] bespoke harmonious reconciliation of sectional differences, racial peace, and a new economic and social order based on industry and scientific, diversified agriculture." From their search would emerge a more progressive South, a landscape of plenty for all, a people eager to embrace the future, not trapped in the past.[108]

Wiggins's unique contribution to this turn-of-the-century intellectual debate was his conviction that forestry was a key to a southern renaissance. It is no surprise, then, that he tried to launch a school of forestry at Sewanee, expecting it would demonstrate to the region how to rebuild its fortunes, environmental and economic. Because funding was unavailable, the project never got off the ground. But Wiggins's ambition to upgrade the university's curriculum by binding it more closely to the contemporary progressive impulse, found expression in his expansion of the medical and law schools, revitalization of the on-site grammar school, and increased investment in undergraduate scientific studies.[109§]

His appreciation of foresters and forestry also impelled Wiggins to embrace a political vision that was at odds with the South's historic disdain for the federal government, a disdain born of slavery, the sectional crisis, and its aftershocks—the Civil War and Reconstruction. As with some of his peers, he realized that the South would rise again only if it were a full, contributing partner in the Union, a point he emphasized in a 1905 article in *Forestry and Irrigation*. Writing about the critical need for a southern Appalachian forest reserve that would stretch from "West Virginia and Virginia through Tennessee and the Carolinas to Georgia and Alabama," he made the case that such a vast national forest would "safeguard the farming, commercial, and manufacturing interests of one of the most important sections of the United States." And only Washington would be able to buy the land and effectively regulate use of its riches, principally timber, water, and coal. The South did not have the capital, expertise, or will to create this much-needed reserve, a point he hammered home in his conclusion: southerners must "entrust the management of this magnificent domain to the wise, liberal, comprehensive administration of the general government."[110]

That was a radical declaration. Few southerners with a memory of the war or its aftermath—and Wiggins, born in South Carolina, had grown up during those years—had ever made so bold. And surely few at the University of the South, that stronghold of Rebel sentiment, were so public in their affirmation of beneficent national governance.

Yet Wiggins's energetic leadership on campus and active promulgation of southern forestry was also in keeping with Sewanee tradition. Like the school's founder, Bishop Leonidas Polk, Wiggins, whose father was a minister and whose father-in-law was Bishop Charles Quintard, believed that the mountaintop campus must exemplify his beloved region's most important cultural values. Like his predecessors, Wiggins was convinced

---

§ Pinchot urged Wiggins not to develop a full-fledged forestry school, instead suggesting a series of courses at Sewanee; "in this way your work at Sewanee would be auxillary to Schenck's [at Biltmore] and [Bernhard] Fernow's [at Cornell]." Wiggins revived the idea in 1903 when Percy Brown, of the Kirby Lumber Company, Houston, wrote suggesting that with the demise of the Cornell University school of forestry, its director, Bernhard Fernow, might be available to head a new school at Sewanee. Brown knew from experience the benefit of well-trained foresters. Kirby Lumber had been one of the first companies to take advantage of Pinchot's *Circular 21* offer to provide private landowners with skilled technical aid; Steen, *U.S. Forest Service*, 54–55. Brown's enthusiasm for forestry was not backed by an offer to bankroll the new school, however, leading Wiggins to respond that "the chief trouble just now would be securing the necessary funds to inaugurate the movement. Without endowment, it is dangerous to undertake too much"; Wiggins to Percy Brown, July 24, 1900, DLA.

that this city on a hill should illuminate the darkness below. So it functioned in this generation, he advised his audience at the American Forest Congress, through its commitment to wise land management. In language that deftly fused Sewanee's religious heritage with its modernist aspirations, the vice chancellor declared that the university served as "a zealous missionary, preaching everywhere and at all times the gospel of forestry."[111]

CHAPTER SEVEN

❧

# *Groves of Academe**

R. Scott Wallinger of Westvaco Corporation spoke for many at a 1991 symposium titled "Forestry Education in the 21st Century" when he vented frustration with the profession's present state and future condition. "It is clear from professional and conservation literature that there is no clear definition of what forestry is today." That lack of clarity had serious implications: "Until we agree on what the term means, until we agree what a forester is, we can't agree on what a forester must know, we can't agree on what a forestry school is, we can't agree on what a forestry curriculum must contain." Rhetorical slights-of-hand would not suffice. "We have symbolically broadened the notion of a forester and forestry by using increasingly the term 'forest resources management,'" Wallinger observed, "but this undefined extension doesn't resolve the issue." No, this confusion over first principles was systemic, a sign of "an antiquated professional structure that no longer serves the profession or the public."[112]

Many among the two hundred researchers, educators, and resource managers at that conference also worried about the broader political ramifications of a profession that seemed in disarray; if foresters did not know how to define themselves, then why should a skeptical public trust them or their forest management practices? The bruising battles over environmental issues during the previous decade—from clearcutting to old-growth preservation—seemed to confirm a marked decline in respect, a sharp loss of authority.[113]

Much of this contemporary handwringing—it continues still—contributes to a narrative of decline. The bleak present (and dreaded future) is set against a glorious past; back then, the argument goes, foresters shared a set of goals, had a precise fix on their social role, and earned a grateful nation's admiration. Although some of this shimmering vision may be true, it is important also to recall how often the profession has been at

* James G. Lewis is the coauthor of this chapter.

odds with itself, how old are the questions that trouble it. There has never been a time when foresters were free from doubt about themselves, their education, or their mission.

In the beginning there was conflict. The three men who were central to the early history of forestry education—Bernhard Fernow, Gifford Pinchot, and Carl Schenck—had decidedly different views on how to establish professional schooling in America. Each believed his perspective was correct and sought through the establishment of competing schools—Cornell, Yale, and Biltmore, respectively—to institutionalize their convictions; that they therefore staked their reputations on the success of these educational ventures only heightened the drama.

The first American-born forester, Gifford Pinchot, found himself in a unique position at a critical moment. He had taken up the profession at the urging of his father before there was even a demand for foresters. There were no forests under practical management, no forestry schools in North America, and the tiny forest preservation movement—if such it could be called—was divided over how to slow the industrial assault about the nation's forested estate. After graduating from Yale College in 1889, Pinchot studied at L'Ecole Nationale Forestière in Nancy, France, and then served a succession of supervised apprenticeships in Germany and Switzerland, where he learned some of the profession's scientific language and technical methodology. He also concluded that European forestry could not be transplanted wholesale to his native country—it was "chiefly valuable as a sort of guide in the study of new conditions and the devising of new methods" better suited to New World conditions.[114]

In subsequent years he would apply another critical lesson gained overseas. A leading Swiss forester had advised the young man to "Go slow with [the] f[orest] org[anization] in America. First mark out state forest[s], then protect them, then [establish a] forest school." This advice reads like a historical summary of the profession's first years. Though the first national forests were created in 1891 just after Pinchot's return from Europe, as secretary of the National Forest Commission in 1897 he helped establish new ones and worked for the passage of the Forest Management Act to bring the forests under scientific management. Three years later Pinchot and his family gave Yale its School of Forestry.[115]

Before Yale, Pinchot had informally taught the principles of forestry to some of the employees working at Biltmore. Others interested in this new profession, including his friend and fellow Yale graduate Henry Graves,

Field training at the Yale Forest School's summer camp at Grey Towers covered many contingents a forest ranger might face, including bear attacks.

came to the estate to learn about American woodlands before heading to Germany for formal training. When Pinchot left Biltmore in 1895, Dr. Carl A. Schenck, a young German forester educated at the University of Giessen, took his place and taught forestry to his apprentices. Two years later, in 1898, with George Vanderbilt's permission and Pinchot's blessing, he formally established the Biltmore Forest School, essentially a one-man operation patterned after the German master schools and intended to appeal to the sons of lumbermen or landowners who wanted a quick overview. The instruction in the one-year program was practical as well as theoretical, emphasizing the management of private forest properties.[116]

It was on pedagogy that Schenck and Pinchot most sharply disagreed. Schenck's curriculum gave his students extensive real-world experience, and this meshed with his goals for their careers. "I had advised my graduates to seek employment with the large landowners of timberlands rather than with the Bureau of Forestry in Washington, because I wanted them to be foresters in the woods rather than foresters in office buildings," he wrote. Openly critical of Yale's more theoretical approach, which he believed left its students ill prepared for practical forestry, Schenck also challenged what he perceived as Yale's anti-industry bias. Lumbering, the head of Biltmore affirmed, "was an essential part of forestry and an integral part of the studies and the lectures offered at any forest school." Any

Carl Schenck (center, in white shirt) in front of his school house with his students. As director of the Biltmore Forest School in Asheville, North Carolina, he clashed with Gifford Pinchot over pedagogy.

school but Yale, that is. When Pinchot learned that "in the school examinations at Biltmore a knowledge of logging and lumbering were weighed higher than that of silviculture or any other branch of 'scientific' forestry," he reportedly called Schenck "an antichrist."[117]

However unlikely he was to cast such a slur, Pinchot urged Vanderbilt to shut down the Biltmore school. That would not occur for another decade, but market forces, not personal animosities, led to its demise. Schenck's nondegree-granting school, with slim resources and informal instructional methods, could not compete with the then-fourteen university-based undergraduate and graduate programs or with the academic credentials they could bestow. Although these more powerful organizations had not resolved "the proper balance between academic and practical training" that Schenck thought so vital to successful professional education, by every other standard they had superseded Biltmore.[118]

Bernhard Fernow had been more adept at forecasting the university's absorption of forestry. Indeed, he initiated the process in 1898 when he resigned as chief of the Division of Forestry to launch the New York State College of Forestry at Cornell, which he saw as an opportunity to build forestry from the ground up. Such was not to be. Fernow had believed the Cornell program would succeed because the 1897 Forest Management Act created a demand for trained civilian foresters, but his efforts were stymied by underfunding, poor facilities, mismanagement, and just plain bad luck. Cornell may have collapsed ignominiously after only five years of operation, but it had considerable impact on the future of forestry education. Most notable in this respect was that Cornell's curriculum established a benchmark by which other forestry schools measured themselves, as did Fernow's *The Economics of Forestry* (1902), and taken together they established the "real beginning of the standardized curriculum of American instruction in forestry."[119]

Much of what the Cornell program had lacked in 1898—funding and facilities—Yale obtained at the outset: in February 1900, the Pinchots gave the university $150,000 to establish the School of Forestry; its two-year master's program was the first of its kind in the country, and Henry Graves, then serving as Pinchot's second-in-command at the Bureau of Forestry, agreed to serve as its dean. In that capacity, Graves developed a curriculum for Yale that built on students' undergraduate work in the general sciences, preparing them for employment in governmental forestry or private industry.

That Graves and Pinchot, two "enthusiastic Yale men," wanted their alma mater to supplant Fernow and Cornell is clear. It would do so by reclaiming American forestry from foreign interlopers: "Fernow was bearing too close to the German pattern of a forest school and to teaching German methods of silviculture," Graves later wrote. Pinchot was just as blunt about Cornell *and* Biltmore: "We had small confidence in the leadership of Dr. Fernow or Dr. Schenck. We distrusted them and their German lack of faith in American Forestry. What we wanted was American foresters trained by Americans in American ways for the work ahead in American forests." Yale's influence quickly spread throughout the emerging forestry profession. Most of its graduates entered government work, on either the state or the federal level; the first five chief foresters of the Forest Service were either faculty members or graduates of Yale's School of Forestry. In short order, it had become dominant.[120]

Yale's success, combined with the implementation of federal, state, and private forestry, bred competition—there was an increased demand for additional schools, especially at land-grant institutions. Their subsequent proliferation created several problems, however. The varied quality of education and lack of standards became the most critical threat to the future of forestry education. In 1909, Pinchot organized a conference "to

USDA FOREST SERVICE

Henry Graves left the Division of Forestry to start the Yale Forestry School in 1900, where he served as dean for nearly thirty years. Graves (back row, second from left) taught three future chiefs of the USDA Forest Service: William Greeley, Robert Stuart, and Ferdinand Silcox (second row, second from right). Several other students went on to establish forestry schools around the country.

consider the aim, scope, grade and length of [forestry] curriculum," and two years later its evaluation, entitled "Standardization of Instruction of Forestry," appeared in *Forestry Quarterly*. This report asserted that professional "training must include a substantial general education, as well as a well rounded course in all branches of technical forestry, and that the standard must be high. Emphasis was placed on a training that would create a body of professional men who could formulate principles and do the constructive work required to put them into operation."[121]

That this vision squared with the Forest Service's managerial model—giving rangers and forest supervisors considerable autonomy in decision-making—suggests how influential the agency's needs had been in curriculum planning. None of the schools of forestry were required to adopt the proposed curriculum, but their cordial response to it helped lay the foundation for a more consistent form of education over time.[122]

Attaining the desired consistency did not stop professional arguments over the proper nature and quality of forestry educational offerings. On the contrary, success bred discontent. Except for a hiatus during the Great

War, enrollments increased throughout the 1910s and 1920s, new schools were created, and a wider range of specializations emerged to complicate the educational agenda of the professional programs. These changes generated a wave of critical commentary on the presumed deficiencies in forestry education. "In addition to concerns over curriculum content and sufficiency," Richard Skok has observed, "writers were equally critical of the inadequate number of well qualified faculty at many schools, the emphasis schools were placing on research, and the inadequacies of facilities." As the drumbeat grew louder, forestry educators did what they have done ever since—they called for another comprehensive study. Henry Graves complied, offering in 1928 a close analysis of the social influences and scientific forces affecting the profession, providing insight into the increased need for both specialized training and natural science coursework, and recommending that an even more complete assessment was necessary.[123]

Graves took his own advice and, with Cedric Hay Guise, produced the single most comprehensive assessment of the profession theretofore. Their book-length study, *Forest Education* (1932), revolved around the by-now standard question: What is a forester? This had no easy answer, and any response was complicated by employers' sharp criticism of the educational attainments of entry-level foresters. To meet employers' pressing needs, the authors proposed—and a series of followup studies concurred— that the Society of American Foresters accredit forestry programs. Controlling the quality of membership, SAF President H. H. Chapman later asserted, was the only way to distinguish "between a profession and a craft," and it was as a profession that foresters had the best chance to control their destiny.[124]

His argument assumed that there was a single definition of *forester* and what he did. That is why he later opposed, for example, a broader interpretation of the term to encompass those who administered "wild" land, for this would have diluted the profession's occupational homogeneity. But not all foresters agreed. Forest Service scientist H. T. Gisborne did "not believe that our Society [of American Foresters] can continue to thrive or that commercial timber growing can even exist on any appreciable area without sharing *equitably* with other wildland uses both the responsibilities and costs of operation." Foresters, in other words, must secure the broadest possible training, and their professional society must embrace as wide a range of occupations as possible.[125]

Chapman and Gisborne represent the conservative and liberal

positions adopted over accreditation issues and the core of forestry education: was forestry mainly timber management, or was the forest to be managed for multiple purposes? The debates continued for decades, with the conservatives holding a comfortable majority. In the late 1930s an SAF questionnaire to its members implied that "real foresters" practiced silviculture, prompting a threatened mass resignation of all foresters in the Department of Interior. In the 1940s many foresters with grasslands responsibilities felt more comfortable in the new Society of Range Management and resigned from SAF. In 1949 the SAF membership voted 2,036 to 1,506 against broadening forestry curricula, favoring the existing focus on silviculture and timber management. In fact, silviculture was the field of study that made forestry "unique," and to dilute that would be unacceptable, as Myron Krueger argued in the *Journal of Forestry*: We can "be experts in making forest lands yield timber crops up to the full potential of those lands to produce, or we can make it a society of specialists covering a range so wide that the general public is confused as to what a forester really is." Hewing to a narrow self-definition was the preferred means to professional survival.[126]

Doing so might also be a means to professional stagnation, worried Gifford Pinchot. The fourth and final edition of his primer, *The Training of a Forester* (1937), reflected his willingness to redefine forestry to remain socially relevant and scientifically current. Unlike earlier editions, the 1937 version revealed significant alterations in his thinking about forests and forestry; ecological insights began to replace utilitarian perceptions. The book's original emphasis on silvics, forest economy, and lumbering was modified through the insertion of material he labeled forest ecology; where once he addressed trees' "individual habits of growth and life," now there was considerable new material on them as "members of plant communities" rooted in diverse climates, physiography, and soils. Trees were not the sole concern, either: foresters must have a good understanding of entomology and wildlife, an approach that suggested that human needs were not always paramount and had implications for forestry as traditionally defined. Foresters "must know about these elements of the forest and their behavior" because they "must work toward maintaining the balance of nature." Foresters would also have to adapt to competing human claims on the environment. To prepare their students for these many challenges, forestry educators needed to teach a more sophisticated form of land management.[127]

They did not take up Pinchot's challenge, but over the next twenty years new voices were raised in support of teaching a more multidisciplinary form of forestry. At the 1947 SAF annual meeting, for example, H. Dean Cochran from the School of Forestry at the University of California at Berkeley argued that the increasing demand to manage forests for their "many and varied uses" required educational curricula that trained professionals more broadly. Concurring was Walter Mulford, head of the Forest Service Division of Personnel Management: "Phases of wildland management other than tree forestry should be recognized as full brothers to tree-forestry in a widely inclusive profession"; similar pleas would be heard through the 1970s. But those who advocated broadening the forestry curriculum found their arguments went for naught. During the succeeding decades, the Forest Service and private companies dedicated themselves to getting out the cut to meet the postwar housing boom and hired silviculturists and road engineers in large numbers. In response, most SAF members favored an accreditation process that ensured continued instruction in the five core fields—silviculture, forest protection, management, utilization, and economics—that Harry Graves had prescribed decades earlier; in this environment, courses on forest ecology and wildlife management were rare commodities. Despite efforts to reform the profession, at the midpoint of the twentieth century forestry remained a highly technical specialty.[128]

That narrow focus would begin to change during the 1960s. Internal and external critics continued to demand the teaching of a more ecologically sensitive forestry, but now their criticisms merged with a pronounced shift in students' interests, which in turn were reflecting the emergence of a powerful environment movement. Educators responded, beginning to market "a variety of courses and curricula in environmental fields," including forestry. The discipline was "sufficiently environmentally oriented to attract a large number of majors as well as masses of students from other majors who were seeking electives with an environmental flavour." Evident in land-grant institutions like the University of Minnesota, Ivy League institutions like Yale, and even the tiny undergraduate-only program at the University of the South, these new course offerings were to provide a more interdisciplinary forestry education, resulting in enhanced employment opportunities. Although researchers concluded that the environmental impulse had left "a permanent mark on forestry education," they also reported that by the late 1970s enrollment pressures had eased, and the "rapid expansion of environmental course offerings had ended." This was perhaps a result of an "environmental backlash" by students and faculty as they realized "the social and economic trade-offs which result from extreme environmental pressures," as well as in industrial hiring practices. The green revolt was short-lived.[129]

Or so it appeared in the late 1970s. The environmental impulse in forestry education, however, proved more durable than its critics then understood. Undergraduate enrollment in forestry programs continued to grow rapidly through the mid-1980s, and graduate enrollment climbed until just before 1990; by the early 1990s, those numbers rose once more. Throughout this period of growth, the marketing of the discipline to a more ecologically conscious student body kept pace. The number of environmentally based courses increased to the point that many formerly old-line "forestry" colleges, departments, or programs had begun to add "environmental" to their names or use the seemingly more neutral moniker "natural resources": Yale's became the School of Forestry and Environmental Studies (1972), for example, and Duke's underwent a similar alteration, emerging as the School of the Environment in 1991. Such shifts in nomenclature signaled the symbolic and real changes occurring in the schools' curricula and academic orientation.

The value of those changes remains hotly debated and continually reassessed. Every few years a new study has taken the profession's pulse, examined the ever-more varied roles that foresters have filled, agonized over accreditation of forestry schools, and concluded that "the term forester has lost much of its former meaning." Inevitably, this seeming loss of definition has led to a dire conclusion—the profession was (and is) in the midst of an "identity crisis." But is crisis the apt term? The relentless self-questioning has been productive, providing professional educators a much more precise understanding of their field and how it operates within dynamic educational systems and political arenas and in response to shifts in employment patterns. Surely this has been a sign of the discipline's ongoing health, one marker of which is its ability to help succeeding generations of foresters evolve so as to maintain what one industrial employer has argued is the "social license to practice forestry." Renewing that license always has been a necessary part of the forester's social compact and, not surprisingly, has been an essential element in the century-long struggle to define the character and purpose of American forestry education.[130]

CHAPTER EIGHT

⚘

# Grazing Arizona

"All history consists of successive excursions from a single starting point," Aldo Leopold wrote in his essay "Wilderness," a point to which "man returns again and again to organize yet another search for a durable scale of values." Although his reference was to human experience writ large, he could have been speaking of the agency for which he once worked, the Forest Service. Periodically, it has been compelled to reexamine its guiding principles, seeking in new language an old need: to make sense of the present to be better prepared for an unknowable future.[131]

This reexamination was especially intense after the 1980s. That is when, in a delayed response to a remarkable set of federal environmental regulations—including the Wilderness Act (1964), the Wild and Scenic Rivers Act (1968), the National Environmental Policy Act (1970), the Endangered Species Act (1973), and the National Forest Management Act (1976)—that challenged its post–World War II claims to authority over matters of conservation and natural resource development, the Forest Service began to revise the words it had used to define its mission. Whether the rubric was "New Perspectives," "New Forestry," or "Ecosystem Management" mattered less than the intellectual effort and professional energy that brought these innovative concepts to life. For each was an attempt to redefine the agency's land management practices, to make them more consistent with shifting political realities, competing legal demands, and an ever more complicated science of the environment. In this volatile context, for instance, it was no longer possible to promise the production of "a completely stable supply of commodities from public lands," an ambition that a previous generation of federal foresters had rigorously pursued. Such a goal "will never be fully realized given the many natural variables that influence land and resource management," then-Chief Jack Ward Thomas asserted in the mid-1990s; this situation was only compounded by "our collective inability to provide firm, fair, and consistent political direction for federal land management." Nothing more fully captured the paradox within which the agency operated than the title Thomas applied to his musings—"The Instability of Stability."[132]

How to locate a different set of values that might make the future more certain, more predictable, maybe even more stable? Act as Leopold had predicted humans invariably responded when confronted with turbulent times: return to some identifiable past from which to begin anew. Chief Thomas especially pursued this tactic through his frequent evocation of the agency's founder, Gifford Pinchot. In public addresses and internal memoranda early in his tenure, he suggested that the first chief's vigorous articulation of the Forest Service's mission at the beginning of the twentieth century was a model for its behavior at that century's end. In a vivid reminder of the degree to which images of the past can speak to contemporary ideas, Thomas, in closing an important address to the agency's leaders, quoted extensively from Pinchot's autobiography, *Breaking New Ground*. In particular, he fastened on the story the first chief told of how, while riding his horse through Washington's Rock Creek Park, he came to understand that conservation, broadly conceived, was not just about resource management. This flash of inspiration, Pinchot wrote, "was a good deal like coming out of a tunnel. I had been seeing one spot of light ahead. Here, all of a sudden, was a whole landscape.... It took time for me to appreciate that here was the makings of a new policy, not merely nationwide but worldwide in its scope—fundamentally important because it involved not only the welfare but the very existence of men on earth." A similar drama—what Pinchot had described as the "lifting of the curtain on a great new stage"—awaited the late-twentieth-century Forest Service, Thomas concluded. "We have come again to a point that is both an end and a beginning. It is our time. Let's get to work."[133]

Pinchot would have been pleased by this sign of his continued relevance, for he had been intimately involved in the creation of the agency's initial set of core values, and later in the 1920s and 1930s had lashed out at his successors when they drifted away from them. That fifty years after his death he might serve once again as a barometer of the agency's willingness to adapt is a testament of the durability of his ideas. That he had become part of the story foresters passed down from one generation to the next would have pleased him, too, because that was what he once had done: he had inaugurated this form of institutional memory when he crafted narratives designed to sanction the actions of the Forest Service (and of its predecessor, the Bureau of Forestry), thereby claiming a more permanent role for it in the management of public lands and elevating its presence on the American political landscape.[134]

### "A High Old Scrap"

Pinchot particularly liked to talk about his experiences in Arizona, a rough-and-tumble landscape that posed potentially irresolvable issues for any who would develop a national conservation initiative. That, in any event, is how Pinchot narrated the stories he told about his visit to the territory in early summer 1900. He had headed west with Frederick Coville, a biologist in the Department of Agriculture, to assess the impact of sheep grazing on the high country and its watersheds and to determine what, and how serious, the consequences might be for those who lived downstream. The relationship between grazing, forest destruction, and irrigation had become what Pinchot described as "the bloody angle," a tense regional conflict between ranchers and farmers that if it escalated, might derail the implementation of conservative management of natural resources. To defuse these tensions was his and Coville's goal when they joined with Albert Potter, secretary of the eastern division of the Arizona Woolgrowers Association, and Con Bunch of the Salt River (Phoenix) Water Users to tour the affected regions.[135]

But first Pinchot had to endure a number of tests. Ever conscious about what westerners thought of "Eastern tenderfeet," Pinchot "strongly suspected" that Potter led the group through some of the roughest, most inhospitable terrain to take his measure of the federal scientists. The forester affected not to flinch, for instance, when deep in the desert the main water keg and individual canteens bottomed out, and Potter guided them to the area's only water source—a "stagnant pool of terrible green water," complete with "rotting carcasses of cattle that had waded in and drunk until they bogged down and died." A thirsty Pinchot downed that rank fluid. Then there was the arduous climb up the White Mountains, at the conclusion of which lay another challenge: somewhere along an unnamed ridge, "in a trackless forest of Spruce," Potter lost a much-prized knife. Pinchot knew what to do: he bet he could find it. "Potter was sure I couldn't," and so "I took a chance, and by good luck I did find it." Having passed this testosterone check, Pinchot knew good things would come. Retrieving Potter's knife had "fixed my status as a woodsman. Such things have their value. Stories get around. I had to meet the Western men on their own ground or be lost." Sometimes you had to backtrack to make a little progress.[136]

The rest of the tour reinforced that hard-won lesson. As the small expedition worked its way up and down the Mogollon Mesa and investigated the White Mountains and other high-country meadows, its members confirmed that grazing seriously damaged forest biota: "Not only do sheep eat young seedlings, as I proved to my full satisfaction by finding plenty of them bitten off, contrary to the sheepmen's contention, but their innumerable hoofs break and trample seedlings into the ground." Their hoofs also tore up the soil, which rainstorms washed off the "hillsides where it belongs into streams where it does not belong," clogging watercourses and reservoirs. For Pinchot, the conclusion was clear: strict regulation was essential to safeguard the land and ensure the future (and freer) flow of water for irrigation and human consumption. That said, Arizona's forests, which contain "much feed that should not be wasted," nonetheless could be grazed, but not "overstocked"; doing so would destroy the capacity of the forests to regenerate. The conservation of woods and watersheds took precedence. When "young trees are old enough to make it safe, grazing may begin again, but never without careful supervision and control." Through their scientific analysis of the land and its carrying capacity, Pinchot and Coville crafted what they believed was appropriate public policy to govern future use of these important public lands.[137]

A critical political reality also shaped their conclusion that northern and eastern Arizona was not one of those regions from which sheep should be excluded. "In the early days of the grazing trouble," Pinchot later recalled, "when the protection of the public timberlands was a live political issue, we were faced with this simple choice: shut out all grazing and lose the Forest Reserves, or let stock in under control and save the Reserves for the Nation." Within this construct, he and Coville acted on behalf of what they conceived to be the greater good and took a crucial step in winning broad public support for the initial management and later expansion of the national forests. In his experiences in Arizona, Pinchot saw the Forest Service's future.[138]

### Moving On

That future had, by the late twentieth century, become the agency's present: in 1900, as a traveling Pinchot platted the intersection between upland grazing, flatland irrigation, and urban developmental pressures,

The decision by the Forest Service to allow sheep on the public rangelands sparked controversy that lasted for years, and provided a management test for the young agency. Here, a ranger from the Wasatch National Forest in Utah is seen counting sheep as they enter the range in 1914.

he identified many of the issues that have come to dominate the modern Southwest—and will continue to define life in the oft-arid landscape. The wider region, running from Texas to California, experienced such enormous population growth after World War II that by 2000 it was home to six of the nation's nine largest cities—Houston, Dallas, San Antonio, Phoenix, San Diego, and Los Angeles; not far behind were Las Vegas, Salt Lake City, Denver, and Tucson.

As a result, urban Arizona has encroached on once-distant wildlands, generating a host of interrelated problems, including increased fire dangers and accelerated threats to critical habitats, watersheds, and rivers. Moreover, the demand for the preservation of natural beauty is forever tangled up with intense desire for mountain and desert recreational space. Slaking the great thirst of the citizens of Flagstaff, Phoenix, and Tucson only adds to the pressures on regional water supplies, already stretched thin meeting the needs of sensitive ecosystems, and the escalating demands of computer chip manufacturers, farmers, and ranchers. Although different in degree from the world Pinchot glimpsed during his month-long journey through the territory, Arizona today is not so different in kind.

The public policy dilemmas he faced have continued to reverberate. In 1900, worried that federal forestry would be rendered immaterial by political conflict, he sought leverage to give himself room to maneuver. Conservation—a mechanism for mediating between competing needs and conflicting demands—gave him access to the civic arena, where he could meet with the affected communities, negotiate acceptable management strategies, and accordingly fashion national policy. "There are a great many interests on the National Forests which sometimes conflict a little," he remarked in 1907. "It is often necessary for one man to give a little here, another a little there. But by giving way a little at present they both profit by it a great deal in the end." This art of compromise and commitment to consensus building were essential to ensure the democratic character of resource management.[139]

That faith compelled the agency to respond to public pressure. Upon returning to Washington from his southwestern tour, Pinchot determined that the Bureau of Forestry must have a division of grazing. Its head, he realized, must be someone with considerable grassroots experience, for grazing "is primarily a local question and should always be dealt with on local grounds." It would be unwise to administer the reserves from afar and "under general rules based on theoretical considerations." Scientific analyses of rangeland were essential: "Local rules must be framed to meet local conditions, and they must be modified from time to time as local needs may require." That being the case, Pinchot knew exactly whom to hire—Albert Potter—a decision he never regretted. Potter's "soft, unemphatic, knowledgeable speech, his thorough mastery of his business [and] his intimate acquaintance with the country and its people" had given him "a standing and influence that were remarkable," Pinchot later recalled. Potter "was the cornerstone around which we built the whole structure of grazing control"—built, it should be noted, from the bottom up.[140]

The need to get closer to the land and the communities that depend on it led the late twentieth-century Forest Service to adopt ecosystem management as a guiding principle. Its scientific insights were to enable the agency to manage resource uses in a more careful, thoughtful, and decentralized manner, and to react with greater flexibility to bioregional or local environmental conditions. Ecosystem management held the key, Chief Thomas argued in 1996, to sustaining biological diversity, supporting social

Albert F. Potter with an elk calf in his arms, Teton National Forest, 1918. Potter, according to Chief Gifford Pinchot, "was the cornerstone around which we built the whole structure of grazing control."

and economic development, and "dampening oscillations in forest outputs." This new prescription for management had its share of political consequences, too: thinking like an ecosystem compelled the agency to begin to coordinate its planning with a constellation of forces—congressional conservatives, free-market economists, local environmental activists, and a host of others—who to one extent or another would like to control or transform the agency's behavior and diminish its authority over the public lands. More profound changes would flow from these interactions and new managerial regimes, argued forest policy analysts Hannah Cortner and Anne Motte: "Adopting the ecosystem management paradigm would mean rejecting traditional resource management policies and practices in favor of policies and practices selected primarily for the purpose of sustaining ecosystem health." This in turn would require "extensive social and political changes, ranging from redefinition of the values that define relationships among humans and nature, professions and citizens, and government and citizens, to the creation, reform, or even dismantling of traditional resource management institutions...." Through ecosystem management, the Forest Service might have become an endangered species.[141]

In that tumultuous, even threatening environment, it was understandable if some of the agency's modern leaders occasionally yearned for what they imagined was a simpler moment in the past, a less troubled time when an aggressive Gifford Pinchot dominated the national stage. He and "our predecessors meant the Forest Service to be a guiding beacon for excellence in land management, research, and assistance to others," Chief Thomas averred; "I believe that is our heritage and our destiny." The agency's actions in this regard depended on securing the necessary funding and policy initiatives that would allow it to seize "'the Bully Pulpit' for natural resource management" and execute "a clearly stated national policy." That never happened, and in 1996 Thomas resigned in frustration. His successor, Michael P. Dombeck, entertained the same prospects, believing that the agency could once again "lead by example" and thereby "redeem" its former status as the preeminent conservation agency in the federal government. He, too, would leave office before achieving his goals.[142]

But as Pinchot's experience in Arizona a century ago indicated, the past is a tangled thicket. His hands were never unfettered, his leadership never unquestioned. His decision to admit sheep on the public rangelands under his agency's control—which he knew would set the course for years to come—was not taken lightly. Yet he also embraced a partici-patory form of governance that would lead him to balance, however awkwardly, scientific insights and political exigencies. In acting as an hon-est broker among rival interest groups, he was at once proactive and reactive, assertive and cautious, principled and pragmatic. That durable set of values still makes good sense.

CHAPTER NINE

❦

# *Back to the Garden*

It says something about our culture that a pair of ice cream moguls—Ben and Jerry—could scoop the forestry profession by making *sustainability* a household term. Who in 1989 would have thought a rich, sweet, frozen combination of cream, vanilla, cashews, and Brazil nuts dubbed Rainforest Crunch could be successfully marketed as a means to preserve the folkways and livelihoods of the indigenous peoples of the equatorial woodlands of Brazil? Or that this gastronomic concoction, which annually used in excess of 150 tons of Brazil nuts, all of which were "sustainably harvested from the most biologically diverse regions" of the Amazon, would feed First World consumers hungry for social relevance, environmental justice, and a bit of butterfat?[143]

It says something, too, about the depth of our cultural memory that we may not recognize just how derivative are the arguments of these caring capitalists. Ben Cohen and Jerry Greenfield were not the first to define the intersection between forest management, economic development, and social equity. That honor surely belongs to those who brought the concept of forestry to the United States in the late nineteenth century, individuals who fully appreciated that this science could be a profound agent for political change and social reform. It is with their belief in the socially ameliorative aspects of forestry that this chapter is principally concerned. Offering an intellectual history of sustainable forestry as social uplift, it tracks foresters' faith that through the development of new land management techniques, a better America would emerge wherein some of its most disadvantaged and marginalized citizens would be elevated with the stroke of an ax. As these reformers would discover, however, the path to such sustained change was never easy and only rarely realized.

### In the Beginning

Among those who recognized the impact that forestry might have on the commonweal and on its poor and downtrodden, and who was drawn to the field as a consequence, was Gifford Pinchot. His training in Europe in 1889 and 1890 had stimulated his perception of the profession's democratic potential, in ways negative and positive. In correspondence to his parents and later in his autobiography, *Breaking New Ground*, Pinchot revealed he was stunned by the elitist quality of Prussian forestry, by its practitioners' disdain for the common people. While touring a forest near Neupfalz, for example, he witnessed a scene he never forgot: "the old peasant who rose to his feet from his stone-breaking, as the Oberfoerster came striding along and stood silent, head bent, cap in both hands, while the official stalked by without the slightest sign that he knew the peasant was on earth." In stark contrast were his experiences in Switzerland which, like the United States, "was not an autocracy." Under the tutelage of Forstmeister Meister, who oversaw Zurich's famed Sihlwald forest, the young American (he was not yet twenty-five) came to understand "the qualities a pioneer public forester must have to succeed in a country like ours—practical skill in the woods, business common sense, close touch with public opinion, and an understanding of how and why things get done in government and politics in a democracy."[144]

These formative European experiences came into play two years later when Pinchot worked for one of the nation's wealthiest citizens—George W. Vanderbilt. As he developed a management plan for the forests of the Biltmore Estate, that "magnificent chateau of Indiana limestone," he would later observe that "as a feudal castle it would have been beyond criticism, and perhaps beyond praise. But in the United States of the nineteenth-century and among the one-room cabins of the Appalachian mountaineers, it did not belong." That contrast, he affirmed, was a "devastating commentary on the injustice of concentrated wealth. Even in the early nineties I had a sense enough to see that."[145]

That he was sensitive to some of the gross inequities of the Gilded Age was remarkable given that he, too, had been born into a family of great means and considerable influence. More striking still was the import of another forest management plan he crafted, this one for the indigenous people of the Appalachians—the Cherokee—who labored under even greater disadvantages than the region's poor white population. In early February 1893, Pinchot tramped over a site the Indians owned, "about thirty-three thousand acres of mountain land, almost wholly covered with forest." As he wrote to his father, "Parts of it are finer than any other deciduous woodland I have ever seen, and other parts of it, which

While Gifford Pinchot was working for George Vanderbilt in western North Carolina, he commented on the "injustice of concentrated wealth" that he saw at the Biltmore Estate.

I did not see, are said to be finer still." He was staggered by the size of some of the trees: the chest-high circumference of one chestnut was twenty-four feet, two inches; poplars measured up to twenty-one feet, and a red oak was more than seventeen—"the largest tree of the kind I have ever seen." But disaster threatened this arboreal heaven. A local lumber agent had signed a contract with the Cherokee to cut the vast tract and had been aided in the negotiations by "certain politicians who are anxious to handle the money." Pinchot was worried that this agent and his political contacts, to feather their nests, would clearcut the woods, leading to the Cherokee's impoverishment.[146]

Conceding that "there is a great deal of ripe timber on the land," Pinchot nonetheless considered it "a great pity" that the "rest of the forest should be more or less sacrificed to the removal of the small portion which ought rightly to be cut." Rather than sell the lumber in one fell swoop, "which would of course mean disastrous injury to the forest on account of the way lumbermen do their work," he proposed an alternative that would lead to the "permanent preservation of the forest and the enrichment of the Indians." Drawing on his European training and his recent practical experience at Biltmore, he suggested that the forest be divided roughly into thirty or forty sectors, "in one of which the cutting would be done each year. By the time the last section had been cut over, the younger trees left standing on the first section would be ready for market." But "the success of this plan would depend very largely on the way the timber was handled. That is, extra care would be necessary in felling and getting out the logs, as well as in selecting the trees to fell. But the cost of such extra care, as the experiment at Biltmore has proved, is comparatively slight, while the difference which it makes in the future of the forest is enormous."[147]

Considerable, too, were the potential social ramifications of his plan. If adopted, it would ensure "a constant annual revenue coming in to the Indians," simultaneously enhancing the material life of the tribe and reducing the "tax on the Government for their support." Just as "certain villages in Europe pay all their school and road taxes from the product of their forest, so it seems to me this band of fifteen hundred Indians might go far to pay for the necessary improvements about their village by the rational handling of this magnificent forest." From such an outcome psychological benefits would also flow: the Cherokee would be "elevated by the influence of steady and responsible work," he wrote in the paternal language of nineteenth-century reform. Late-twentieth-century forestry reformers might be offended by his condescension, but they should not mistake the larger thrust of Pinchot's argument. In imagining a scenario in which sustainable forestry, a rising standard of living, and political empowerment were inextricably linked, Pinchot had devised a way by which to enfranchise the Indian peoples of western North Carolina so that they would not resemble a degraded Prussian peasantry.[148]

Nothing came of Pinchot's proposal; there is no evidence that he submitted it to the relevant local or national authorities, and besides, he was outside the system of governance that determined the Cherokee's economic life. But once he had become the chief of the Division of Forestry in 1898, he dusted off his earlier plan and began to articulate a policy in which forestry would grapple with some of the needs of Native Americans. In the late 1890s, for instance, he became deeply involved in the creation of the first Minnesota national forest. As with his earlier scheme in North Carolina, the new forest was designed in part to halt political corruption that had led to the outright theft of Chippewa-owned timber and land and the backroom deals that had robbed the Chippewa of their rightful profits.[149]

Such widespread fraudulence also led Pinchot to seek a closer relationship with the Indian Office in the Department of Interior. In 1908, as chief of the new Forest Service, he forged an alliance with that office, which controlled twelve million acres of forest containing timber whose worth Pinchot estimated was $75,000,000. "No one in the Indian Office or on the ground was capable of handling these forests," he asserted, and the "result was what you might expect." Throughout the nation, Indian peoples "were being cheated right and left by contracts unduly favorable to the purchasers of Indian timber," or by the "failure of Indian Agents to enforce such contracts as they had." In addition, most forests were simply clearcut, making for a tremendous loss of young growth and decreasing the chances of natural regeneration; that there were no provisions for reforestation only made matters worse. But nothing struck him as more absurd and devastating than the story he had heard of an Indian agent who had "sold for lumber the sugar bush upon which his Indians depended for their maple syrup." The Indian Office, he asserted, had no sense of either conservative forestry or the social benefits that accrued from it.[150]

By contrast, Pinchot believed that the Forest Service, which recognized the connection between land management and political reform, was in a position to produce substantial results. Eighteen months after inking a contract with Interior officials to handle the reservations' forests, Pinchot would boast that his agency had "saved large sums of money to the Indians, [given] many of them profitable employment, and by the introduction of Forestry promised to make that employment permanent." These first steps would help those he considered the original conservationists, who once had handled natural resources with "foresight and intelligence," to do so again.[151]

His idea was never fully developed because in 1909, Richard Ballinger, who recently had been appointed secretary of Interior, halted the working arrangement between the Forest Service and the Indian Office; it proved to be one of the sources of the later Ballinger-Pinchot controversy that devastated the Taft administration in 1910. Not for another twenty years, the chief forester believed, would the idea that Indian forests should be "handled not for the profit of political contractors, but for the lasting benefit of the Indians and the rest of us," regain political ascendance.[152]

### Midlife Reflections

Pinchot's belief was not entirely accurate, for in the interim there were important attempts to establish conservation and forestry on reservation lands. A central figure in these initiatives was J. P. Kinney. Trained

Jay P. Kinney, photographed at the time of his graduation from law school in 1908, about two years before he joined the Indian Forest Service. As a "Pinchot man" going to work for the untrustworthy Interior Department, he "had a hard row to hoe for a while."

in forestry at Cornell University and law at National University, he entered the employ of the Indian Forest Service in 1910, at precisely the wrong time. Caught in the crossfire between the Department of Interior, in which the Indian Forest Service was housed, and the Forest Service, Kinney found his first years extremely exacting. Interior employees, he later recalled, thought that because he was a forester, he must be a "Pinchot man"; those in the Forest Service considered him untrustworthy because of his employment in Interior. Collateral controversies within Interior further confounded Kinney's labors: "I had a hard row to hoe for a while."[153]

He plowed on. Although he believed "the young fellows in the Forest Service were overidealistic," he shared their critique of his Interior predecessors' actions: "Soon after I began with the Indian Service," he remembered, "I learned that the things that had been done on the Indian lands were not wise. Therefore, my sympathies were with the Forest Service, as far as forestry went." That led him to espouse partial cutting to produce a steady stream of revenue for the tribes. On the Menominee reservation in Wisconsin, on the Klamath in Oregon, and throughout the West, the understaffed and underbudgeted agency sought to make the case that clearcutting forests was bad science, poor economics, and misguided social policy. "The attitude of the lumber industry in Wisconsin, as well as that of the Menominee Indians, was very hostile to the introduction of lumbering methods involving the expenditure of funds for future forests if the effect was to reduce current income." A turning point came, he believed, when he learned that a local, white-owned lumber operation had been conservatively and profitably harvesting its lands for some time. "I used the fact that the Goodman Lumber Company was adopting selective cutting to convince the Menominee Indians that it was practical."[154]

His claim only partly convinces. The Menominee had a pragmatic set of reasons of their own for pursuing this "new" cutting strategy, and it predates Kinney's assertions by many years. In the 1860s, convinced that white timber interests, known locally as the Pine Ring, were illegally and destructively harvesting on the reservation, the tribe established a committee to defend its interests; it regularly filed complaints with the Department of Interior about the clearcut depredations and subsequent loss of income. Seeking to retain control of the resources on its property and to develop an internal skilled labor force, the Menominee constructed a sawmill and lumber camp in the early 1870s; from this enterprise, the tribe gained considerable experience and financial reward. Its bid for self-sufficiency in this instance was temporary—the federal government ordered logging to cease in 1878—but it was exactly this kind of outcome that led to continued tribal agitation in the late nineteenth and early twentieth centuries for greater control over its lands and their productive capacity. In response to these demands, the La Follette Act of 1908 granted the Menominee the right to commercially log and mill timber, under the supervision of the Forest Service. In arguing on behalf of the bill that would bear his name, Wisconsin Senator Robert La Follette waxed enthusiastic: "The forest is the natural home of these men. They are what is known as 'Timber Indians.' Their every instinct teaches them to seek a livelihood from within the forest." That instinctive response would be guided by a desire to ensure a sustained yield. In these northern woods, "the harvest of the crop of forest products should be made in such a way that the forest will perpetuate itself; that it shall remain as a rich heritage to these people from which, through their own labor, they may derive

A Forest Service logging crew, consisting mostly of Menominee Indians, at work on the Menominee Indian Reservation, Neopit, Wisconsin, 1909. The tribe's interest in protecting their forests dated back to the 1860s.

their own support, and that, too, without ruthless destruction." His words spoke for many though not all Menominee, a convergence that suggests that when Kinney voiced similar arguments years later, his words fell on already receptive ears.[155]

That Kinney's interpretation of events framed the Menominee as a naive people forever acted upon by more sophisticated whites suggests the degree to which his perspectives on the prospects of American Indians diverged from those of other foresters. He allowed that he was considerably more skeptical than visionaries like Pinchot and La Follette, who he felt overestimated the noble character of American Indians and overemphasized the role the federal government should play in their restoration. In *A Continent Lost—A Civilization Won* (1937), *Indian Forest and Range* (1950), and a memoir, *My First Ninety-Five Years* (1972), he proclaimed the virtues of complete assimilation of Indian peoples into the dominant white culture. His decades in the Indian Forest Service, he observed in his autobiography, had led him to "the conclusion that the nourishing of the idea or notion in the mind of the Indians that they should remain an insulated group, separate from other inhabitants of the United States, constituted the greatest obstacle to their social and economic advancement." Disputing the well-established record that by hook, crook, and purchase whites had absorbed prime Indian lands, and believing that such transfers were not necessarily bad in any event because they forced Native Americans to come into greater contact with prevailing white cultural values and social norms, Kinney insisted that the reservations were traps. Never snared like those do-gooders whom he brushed off as "short haired women and long haired men," he wrote *A Continent Lost—A Civilization Won* in eager rebuttal of what he described as the "urgent propaganda" of one of these starry-eyed idealists, his superior in the agency, John Collier.[156]

Collier, whom President Franklin Roosevelt had tapped to be the commissioner of Indian Affairs in 1933, apparently was unruffled by his subordinate's contrariness; Kinney remembered showing his boss the manuscript for *A Continent Lost—A Civilization Won*, and Collier merely shrugged, suggesting "'Go ahead and publish it.'" The commissioner was unfazed because, like Pinchot before him, he was convinced that more needed to be done for the Indian peoples through a renewed emphasis on land reform and conservative forestry. The link with Pinchot is not gratuitous; Kinney, for one, believed the two men were similar types of crusaders.[157]

The campaign Collier launched became known as the Indian New Deal, the essential characteristics of which he sketched out in the initial heady days of Roosevelt's first term. In collaboration with his special adviser, forester Ward Shepard, and Robert Marshall, then chief of the Indian Forest Service, Collier outlined a bold plan to alter the status of the Native American peoples. They went public with their ideas in an article in the *Journal of Forestry*, the title of which—"The Indians and Their Lands"— captured the authors' desire to reestablish Indian rights to indigenous grounds. As the authors acknowledged, the "Indian forest problem is only one phase, though an extremely important one, of the whole Indian land problem." At its heart lay the Allotment Law of 1887, which had commanded that tribal lands be distributed to individual members of the community so as to transform "the Indian into a responsible, independent, self-supporting American citizen by the over-simple expedient of mandatorily applying to him the individualistic land tenure of the nineteenth-century white American." The consequences were devastating, indeed the exact opposite of the stated purpose of the allotment law: with the loss of more than sixty-three million acres, "much of it the best" once under their control, the Indian peoples had become "landless," deprived "in large measure of their chief means of support without substituting any other means in its place." A miserable failure, the allotment policy demonstrated just how "dangerous it is to try to solve problems by theories not soundly based on the facts of life and nature."[158]

Repairing this damage required a new approach and a different set of assumptions. Collier, Shepard, and Marshall proposed to consolidate or restore as much land as possible to communal ownership, utilizing land exchanges, purchase, and where possible, relinquishment of allotments. This dramatic shift in land tenure on the reservations would be combined with ongoing training in the management of forested lands. The authors envisioned a harvesting system much as Pinchot had forty years earlier, in which "a light selection method of cutting" would be employed, one that would remove "not more than fifty per cent of the volume of the stand." This would leave "sufficient growing stock to make it profitable to return for at least one and perhaps several additional cuttings before the end of the rotation." And again like Pinchot, they believed that such a logging strategy would work for the forest and for the people who depended on it: "The operation will…bring to the Indians the power to manage their own affairs and the self-respect which such power insures."[159]

A similar restorative impulse was manifest in new arguments emanating from the Forest Service about its future course. This intellectual connection was not surprising, given that Collier's coauthors had been members of the Forest Service and were friends with Ferdinand A. Silcox, the new and fifth head of the agency. Silcox believed that the forestry profession of the mid-1930s was at a critical juncture. He urged his colleagues to recognize that "[w]e must fit forestry into new economic and social conditions. Twenty-five years ago, relatively few persons could foresee the consequences of unbridled exploitation and over-development of all resources. Now these consequences are so clearly evident that few deny them." What that evidence of environmental devastation had clarified was the need for a new philosophical orientation that rejected "rugged individualism" because it often gave license to "the strong to take advantage of the weak." Instead, arguing that forestry ought to be "an instrument for social and economic betterment," he proposed basing "our forest policy not merely upon the need for timber, but also upon such other considerations as stability of communities and employment, dovetailing of agriculture and forestry, and balanced use of land resources."[160]

His articulation of forestry's communal responsibilities, paired with his call for the profession to return to its former "crusading spirit of translating forestry ideals into actual life," won the praise of only the small coterie of socially minded foresters who embraced Silcox's notion of forestry as "social service" and accepted Collier's assumption that "intelligent, permanent land use" and "human understanding" were crucial to the resolution of the nation's ruinous Indian policy. But many more foresters have come to share this faith and consciously or otherwise have incorporated the earlier generation's arguments about economic opportunity and social regeneration into today's sustainable forestry.

"The definition of sustainable forest management that is now evolving," V. Alaric Sample and Roger A. Sedjo noted in 1996, "requires meeting three conditions simultaneously; it must be ecologically sound, economically viable, and socially responsible"—a triple bottom line. This blend is essential to ensure the success of a more ecosystemic form of land management and reflects "a difficult lesson" that environmentalists learned at the end of the twentieth century in "developing countries around the world"—that it is impossible to secure "long-term protection of forest ecosystems without incorporating the economic and social needs of the local people into conservation strategies." Forestry and foresters must be as concerned with sustaining the land as with the development of "sustainable communities," the two being parts of a whole.[161]

That reciprocal characteristic of sustainability is vividly captured in the ten principles and criteria that the Forest Stewardship Council adopted in August 1994 and updated since. Founded a year earlier, FSC emerged as an independent and international organization whose membership has included a broad spectrum of interest groups—environmentalists and foresters, representatives from indigenous peoples' organizations and timber companies, as well as those involved in forest products certification. To secure FSC's independent certification of forest products as grown on lands managed according to a set of environmental, social, and economic standards, producers must adopt and demonstrate their adherence to the prescribed rules. This entails, for example, complying with all applicable national laws and international regulations, establishing "long-term tenure and use rights" to the affected land, enhancing "long-term social and economic well-being of forest workers and communities," conserving biological diversity, and maintaining sites of "major environmental, social, or cultural significance." Of particular note, given some foresters' concerns in the past about links between exploitative commercial development and social oppression, is FSC's third principle, concerning the indigenous peoples' rights. Those organizations desiring FSC sanction must recognize and respect the "legal and customary rights of indigenous peoples to own, use and manage their lands, territories, and resources...." And they must as well accept that indigenous peoples "shall control forest management on their lands and territories unless they delegate control with free and informed consent to other agencies." Additional constraints involve the adoption of a forest management regime that "shall not threaten or diminish, either directly or indirectly, the resources or tenure rights of indigenous peoples," protects sites "of special cultural, ecological, economic or religious significance," and compensates indigenous peoples "for the application of their traditional knowledge regarding the use of forest species or management systems in forest operations." Through economic incentives and moral suasion, FSC has sought to empower historically disadvantaged peoples, restore devastated woodlands, and develop a greener marketplace for forest resources.[162]

Judging by the increased support that FSC principles—and others like them—have gained from among forest resource professionals in academia, industry, and government, sustainable forestry appears to have entered a new stage in its development. That is in part because it has become a matter of international significance, a logical outcome of the location of many of the affected forests—in the tropics and within Third World nations. Consistent with this globalization of the idea of sustainable forestry is the growing concern for the maintenance of indigenous peoples and the rainforests within which they live. These worries, echoing late-nineteenth-century concerns about the disadvantaged, revolve around the use of economic systems and political reforms to establish conservative land management policies to enhance aboriginal power within an increasingly global timber market.

Although the First World has much to teach the Third, the reverse is equally true. Most "Amazonian Indians continue to be like their forebears—they are still Indians—certainly in terms of their plant resources and the ways in which they use and manage these," William Balée has observed. But intriguingly, he compared their land management techniques with those of the developed world: "The resource management practices of the indigenous farmers and foragers of Amazonia of today are less destructive of the environment, by any measure, than our rapacious nation-states with economies based on the burning of fossil fuels." Those with few resources and little power are not responsible for the industrial poisoning of rivers and lakes; they are not complicit in the "increasingly apparent scenario of major biotic depletions." That being the case, Balée and others have suggested that First World conservationists must "rethink their premises" and recognize that their drive to protect Amazonian lands and peoples has two purposes—to ensure the maintenance of the rainforest biota and human ecology, and to increase their own cultures' chances for survival. "If modern states cannot protect the remaining Indian villages and non-state societies of the world," Balée concludes, "will they ever be able to emulate them in terms of resource management, and biological and ecological diversification?" Sustainable forestry will save us all.[163]

What accounted for the greater receptivity to this claim of salvation in fin-de-siècle America? There were myriad factors, perhaps the most prominent of which was the rise of environmentalism in the United States since the early 1960s. A social movement and cultural critique that challenged "the dominant, development-oriented current of post-war American society," in historian Hal Rothman's words, it emerged as a powerful force in an increasingly affluent culture in which "quality of life" issues came to define the political landscape. Strengthened by certain seminal texts—Rachel Carson's *Silent Spring* and Paul Ehrlich's *The Population Bomb*—and reinforced through battles over natural landmarks such as Dinosaur National Monument, the movement pressured Congress to enact stronger environmental laws governing wilderness, environmental land management, water quality and quantity, and a host of related initiatives; their passage has transformed the way many Americans have come to think about their place in the world.[164]

That environmentalism has led to intense political struggles over, even gridlock on, pressing land management issues was all too evident at the close of the twentieth century. Surely this was yet another compelling reason why contemporary foresters—a large cohort of whom had come of age in this environmental era—were drawn to idea of sustainability. For them, it represented an enticing middle ground on which competing forces could meet to discuss and perhaps resolve any number of essential, if occasionally contradictory, needs. Some of what drove "advocates of sustainable development," as William DeBuys pointed out, was their faith "that economic use of environmental resources can be made compatible with good stewardship of them," and their belief that this balance could be maintained over the long run. Sustainability, and its attendant language of consensus, was also alluring amid the often-brutal rhetorical clashes over wilderness values, economic development, and social justice. Speaking to this felt need to locate a space in which the combatants could more safely meet, argue their differences, and resolve some disagreements was the motto of the Seventh American Forest Congress (1996): "Many Voices…A Common Vision." That through sustainability we might reach the promised land of conflict resolution was made all the more delicious knowing we could accelerate our arrival there simply by purchasing a pint of Rainforest Crunch.[165]

CHAPTER TEN

❦

# *Logjam*

It was a bit of eavesdropping that gave me my first clue as to why the Seventh American Forest Congress, held in late February 1996, had been convened. Seated in a cramped, darkened bus slowly rolling through the sodden Virginia countryside on our way from Dulles airport to Washington, D.C., site of the meeting, I overheard snatches of a conversation: "...when ecological devastation threatens," a deep yet quiet male voice murmured from behind me, "we just have to pull together...." What caught my attention was the emphatic "just *have* to," implying that a social responsibility that ought to lead people to protect stressed ecosystems was routinely ignored: a critical obligation was left unmet, perhaps even unrecognized.

That plea set an appropriate tone for a gathering that expected to formulate the nation's future forest policies, and then to lay down ground rules by which to meet the obligations these policies would impose. The timing could not have been more opportune. As the voice I had overheard had suggested, the conflicts over policies governing the management of public and private forested lands had been intensifying since the early 1990s. This had been as true for the Intermountain West and the Pacific Northwest as for the southern Appalachians and the Adirondacks. Some of these disputes were local in origin and significance, but others were national, most notably the furor that erupted over the so-called Salvage Rider. Attached to the 1995 Rescission Bill, the rider had suspended environmental legislation and upped the cut on the national forests. This piece of "stealth legislation," as the *Seattle Post-Intelligencer* called it, which was slipped into law without public notification or hearings, not only infuriated environmental activists but further undercut the prospects of engaging competing stakeholders in America's forests in a civil dialogue. That was strikingly evident in the reason given by Mike Roselle, cofounder of *Earth First!*, for boycotting the congress: "You don't sit down with lumberjacks," he told the Associated Press, "to decide what to do with the forest."

Offering a respite from this "painful and perpetual tug-of-war" is what initially had led the organizers of the forest congress to announce the three-and-a-half-day assembly, hoping it would ease tensions and temper the rhetoric. Once quieted down, the energies formerly consumed brawling over national forest policies could be focused instead on establishing a new basis for consensus. Out of our "many voices," the congress motto prayed, would come "a common vision."

I was not convinced that this prayer for comity would carry the day, especially when, a month before the conference, a letter arrived from its board of directors and senior sponsors—a remarkably diverse lot, ranging from environmental leaders to industrial foresters, from Brock Evans, formerly of the Audubon Society, to Sharon Haines of International Paper. In their missive, the sponsors sought to allay fears that the congress was "a front for any special interest or political agenda" and promise that it was not "being rigged by the 'timber beasts,' or the 'radical enviros,' or the 'pointy head academics and bureaucrats.'" The letter reaffirmed that the meeting's stated purpose—"to develop a process that encourages new and more open communication between interested parties about future management and policies governing our nation's forests, both public and private"—was its true intent. This reaffirmation notwithstanding, it was clear that the congress would not be free from the rhetorical posturing that has so polarized the very forestry issues it wished to resolve.

To finesse the disruptive pattern of negotiations in which, as Yale forestry professor John Gordon observed, "everybody shouts at each other until they're tired and then goes to court," the congress established a unique format to compel meaningful dialogue: the approximately fourteen hundred participants were seated ten per table, the demographics of which were as diverse as the coordinators could make them. Those of us seated at Table 74, for example, came from all over the country, worked in public and private forestry, were involved in the environmental movement, or were students and teachers interested in these issues as they influenced urban lives and rural landscapes. This representative mix of backgrounds and interests might have confounded our capacity to devise vision statements, develop principles by which these visions would be framed, and delineate some practical first steps to achieve these goals. Because our intense work was conducted almost exclusively at the tables, however, we were forced to talk *and* listen, to argue *and* negotiate, and to do so face-to-face, all within the tight confines of a round table. The physical setting and the length of time we sat together had a subtle psychological impact:

bound to and within that closed circle, our commitments and conversation increasingly focused on this microcosmic realm, a bond that for the most part facilitated earnest discussion and civil exchange between people who, in another context, might well dismiss one another's perspectives. It's hard to turn your back on people when you are looking into their eyes.

Not that participants didn't try to stare one another down. In such close quarters, with the tables pressed in on one another, it was easy enough to pick up raised voices, sharp exchanges, turf fights. These flareups found their most visible expression when on the first full day of the congress, Steve Kelley, an environmentalist from the Flathead region of Montana, strode to the main podium, seized a microphone, and announced that he and other environmentalists were "tired" of being ignored, of having their issues swept aside; he called an ad hoc press conference to air their disappointments. His proclamation was greeted with catcalls and jeers and a few cheers. "Put your agenda down on the table," one voice boomed. Kelley had stepped up on the podium, his amplified voice had reached all the conferees, and he had called upon like-minded souls to abandon their close-knit conversations and to coalesce around a different set of concerns in another room with the press as audience—a move that challenged even as it acknowledged the power of table talk.

The table format may have also accounted for the general lack of controversy over the particulars of the final versions of the "vision elements" of the congress. Those that passed most readily were crafted to secure support as broad as possible. The first element is a case in point: "In the future our forests…will be maintained and enhanced across the landscape, expanding through reforestation and restoration where ecologically, economically, and culturally appropriate, in order to meet the needs of an expanding population." Not surprisingly, eighty-five percent of the delegates affirmed this position, but solid majorities also embraced a call for a nationwide commitment to sustainable forestry that must "support biological diversity" and "maintain ecological and evolutionary processes," and an assertion that future forests would be managed "on the basis of a stewardship ethic with respect, reverence, and humility." Taken together, these vision elements represented a fascinating set of declarations incorporating some of the central principles of cutting-edge environmental science and cultural criticism. Biodiversity and reverence are not the first terms that spring to mind when contemplating the guiding forces in the century-long history of American forestry.

At the time, no one had a clue whether these ideas would influence future legislation governing logging on public and private lands. Still, they went well beyond anything I had anticipated would emerge out of the congress, an anticipation based on the controlled character of earlier forest congresses.

The previous ones, including those held in 1946, 1953, 1963, and 1975, had been in-house affairs, cliquish confabs that brought together a carefully selected list of industrial and public foresters, leaders of the pulp industry, and a handful of kept academics. Their agendas rarely strayed from the subject of timber and took little notice of the rising power of environmentalism. That's how the Seventh American Congress should have been devised, too, its organizers were advised when they visited with German foresters to extend invitations to interested parties worldwide. The forestry faculty at the University of Freiberg, a friend reported, were appalled that nonforesters would be involved in the American congress. "That would never happen here," she said; "foresters never would give up their control." Previous American initiatives shared the German disdain for nonprofessionals and regularly excluded conservationists and the concerned public, and predictably, they produced policy statements consonant with industry's needs. The bottom line drove all.

According to one protester, it did at the seventh congress as well: in its final hour, he circled the assembly holding over his head a sign on which the official motto had been edited to read, "Many Voices…A *Corporate* Vision." This spoke to many environmentalists' deep frustration over the perceived failure of the congress to denounce the Salvage Rider, a failure some felt was due to the influence industrial foresters wielded over the conference agenda and voting procedures. True, a resolution explicitly demanding the rollback of the rider had been overwhelmingly defeated, but another asserting that all national laws, including rules stipulating open deliberation of legislative initiatives—a broader, more inclusive principle—passed easily. The congress, in short, went on record opposing "logging without laws."

I saw in that and other resolutions—particularly those emphasizing ecosystem management and biodiversity—a sea change in the public debate over forest policy in the United States. The first to cast doubt on my optimistic take was a professional environmentalist from Montana with whom I chatted on our way back to Dulles. Rather than achieving their agenda, he demurred, conservationists "were only able to hold the line" during the proceedings. His skepticism suggested more generally that we had not yet located the common footing necessary, or common language required, to pull together.

**Divider photo:** *A film crew readies a wall that will stand in for Yosemite Valley's El Capitan during filming of* Star Trek V: The Final Frontier. (Photograph courtesy of Lisa Strong-Aufhauser, Strong Mountain Productions, Boulder Creek, CA)

❦

# CHAPTER ELEVEN

# *Snapshot, 1897*

"I think my first great realization came through my camera," photographer Edward Weston noted in his daybook. "At least it brought me into closer contact with nature, taught me to observe more carefully, awakened me to something more than casual noting and romantically enjoying." By looking through his lens, Weston believed he could see into his self, a fusion that made him feel, if only ephemerally, at one with the landscape he hoped to capture on film: "Even then I was trying to understand, getting closer, becoming identified with nature. She was then as now, the great stimulus."[166]

Although Gifford Pinchot was no Edward Weston, he shared the artist's belief in the camera's affective power and visual impact. That's why he lugged one with him on an arduous western journey he undertook as confidential forest agent for the Department of Interior during the summer and fall of 1897. Interior Secretary Cornelius Bliss had hired him to evaluate the controversial national forest reserves that President Cleveland had announced, and to report on which of their twenty-plus million acres should stay within the emerging system, and which should be returned to the public domain. Pinchot, who a year earlier had examined some of these lands as a member of the National Academy of Sciences Forestry Commission, was delighted with his new assignment—it would get him back into the woods, increase his understanding of their complicated political context, and enhance his prospects for creating (and heading) a federal agency to manage these remarkable public lands.

This once-in-a-lifetime experience began in mid-July, when, with his brother Amos, Pinchot took a train west to Blackfoot, Montana; it concluded in mid-November, after an exhausting tour that had carried him through Montana, Idaho, Washington, Oregon, Wyoming, and South Dakota.

Everywhere Pinchot traveled, he snapped away at the landscapes through which he moved. Presumably, he expected these images of badly

Gifford Pinchot took this photo looking north along the main street in the lumber town of Monte Cristo, Washington, in the Cascade Mountains, 1897. The town is located on the Mount Baker-Snoqualmie National Forest. Note the large tree stumps left next to buildings along the street and the wooden barrels on the porch of the tallest building.

burned and grazed lands, soaring forests of hemlock or fir or pine, and stunning panoramas would help convince his audience that forestry and foresters were essential to the conservation of the many natural resources and beauties he identified during his work. That was the strategy he had employed with *Biltmore Forest* (1893), a heavily illustrated pamphlet he published to accompany the Chicago World's Fair exhibit he mounted about his initial forestry endeavors in Asheville, North Carolina. When in 1898 he would enter government employ, Pinchot would make ready use of this modern medium to illustrate his convictions in congressional hearings, departmental publications, and public assemblies. But if he had expected that his 1898 report to Secretary Bliss would be replete with the images he captured in the West, he must have been disappointed. It contained none of his photographs, a remarkable lapse given the striking quality of the exposure he made of the small mining town of Monte Cristo, Snohomish County, Washington.

As he documented his travels, in words and on film, Pinchot noted the variety of challenges that would confront those who might manage the new forest reserves. Not least of these difficulties, to judge from its regular mention throughout his journal, was fire, which ravaged the forests' economic utility and aesthetic value. While camping along Idaho's Priest Lake in mid-July, for instance, Pinchot photographed "[m]uch old burn," and he jotted down that but "for the fires this would be an exceptionally beautiful place."[167]

It was a similar desire to record the human impress on the West that led Pinchot to set up his camera on a hillside overlooking Monte Cristo, a community high on the western slope of the Cascades nestled within what was then the Washington Forest Reserve (now the Mount Baker–Snoqualmie National Forest). Getting there had been difficult. Although a rail line between Monte Cristo and the port town of Everett had been constructed in the early 1890s, shortly after a prospector had tapped into the area's mineral riches, trains did not run every day, much to Pinchot's dismay when he reached the Everett terminus on Friday, August 13. He had to cool his heels until Monday, a delay that allowed him to organize his affairs and settle his accounts, and that also drove the restless thirty-two-year-old forester not a little crazy. By Saturday, a bored Pinchot was forced to browse the hotel's minimal library: "I read until 2 a.m., fool stories of no account in magazines."[168]

He was no doubt relieved to board the 7:45 a.m. Monday morning local to Monte Cristo, which was scheduled to arrive in the tiny settlement in the early afternoon. Within a few hours of his arrival—by 7:00 p.m. in fact—he had slung his twenty-pound pack on his back, and he and Amos had headed south toward Columbia Peak, on their way to Index. After two days of hard hiking in very smoky conditions, they reached their destination and caught a train back to Seattle.

For all the brevity of his stay in Monte Cristo, Pinchot's photograph of it was carefully composed. Its framing device is the clusters of large stumps that fill the foreground, left and right. They anchor the viewer's perspective, and their motif is then repeated in those located farther down the slope, on either side of the street. Their mute testimony to the fierce human energy needed to clear and claim this once-thickly wooded hillside is replicated in the clearcut swath visible on the mountainside that rises to the right.

Just as the many stumps reflect the town's hasty construction—the wooden stores and homes, themselves a product of the fallen timber, are squeezed in between these massive remains—so, too, does the pathway that falls away from the photographer's feet. As it heads downhill, our gaze trips over a tangle of rock and root. With the forest toppled, the erosive force of stormwater would continue to cut through the town's streetscape, though it was the local mine's closing in 1907 that ultimately turned Monte Cristo into a ghost town.[169]

Catching the community in its heyday is not the only significance of Pinchot's photograph. Indeed, its structural elements recall those in the most important painting in his father's Hudson River School collection, Sanford Gifford's "Hunter's Mountain, Twilight" (1866); Gifford grew up with the painting of the Catskill Mountains and would inherit it at James Pinchot's death. The foreground depicts a logged-over declivity through which runs a thin trickle of water that carries the viewer's eye toward a farmhouse barely visible in the shadows; this landscape is also replete with stumps, just as the distant mountain, once cloaked in thick stands of hemlock, has been cut over. That Gifford Pinchot, consciously or otherwise, set up his shot of Monte Cristo in the aesthetic tradition of his godfather (and namesake) reminds us that for him, the ax was double-edged, "a symbol of economic progress and cultural poverty, of conquest and death."[170]

It is unknown when during his five-hour stay in Monte Cristo Pinchot took this photograph, though the shadows angling across the rough-shingled roofs suggest it was shortly after his arrival. Regardless of the timing, when he left town, he would have to walk back over the same elevated vantage point from which he had clicked his shutter, clambering uphill with Amos—who may be the figure stationed in the photograph's middle ground to provide a human dimension—on their way to the mountain pass where they would camp that night, a high country of "good wood and bad water." Pinchot would never return, and so his image of Monte Cristo also marks a temporary intersection between person and place, a visual footprint of his presence in the town's truncated history, a still moment in the rush of the day.[171]

CHAPTER TWELVE

❦

# *Green Screen*

The setting was sublime. With majestic Half Dome dominating the horizon, Vernal and Nevada falls plummeting into the vast floor of the Yosemite Valley, and pine trees jutting up from craggy rock faces, producer Larry Hott and his crew from Florentine Films had no trouble finding stunning footage for their documentary on that most riveting moment in the history of this national park—the early-twentieth-century fight over a dam that would inundate a portion of the Yosemite known as the Hetch Hetchy Valley. But their attention to their work wavered occasionally: as Hott and the others crisscrossed this grand corner of the Sierran landscape, they kept seeing large crowds "pointing at people climbing on the rocks." Finally, their curiosity piqued, they stopped to see what all the fuss was about, only to discover that they had come upon the filming of a scene for *Star Trek V: The Final Frontier* (1989). When they arrived, Mr. Spock was hovering in jet boots near Captain Kirk, who was scaling a rock formation that looked unlike any other the Florentine crew had seen while exploring the valley's rugged terrain. There was a reason why this was so—the Star Trek rocks were fake. The Hollywood film crew had constructed this massive set in the national park to make their backdrop seem real, Hott observed, "and it made me think that perhaps in the future all we would have was a memory of the way the wilderness actually looked. [Then] we will have to reconstruct it," just as the film crew had done.[172]

This anxious speculation about the future of nature, and our place within an increasingly "artificial" environment, is part of a larger cultural debate that since World War II has helped define the subject and thrust of documentary films about nature. Also implicit in this debate, as it is in Hott's commentary, is that nonfiction films or documentaries—in contrast to Hollywood blockbusters—are more genuine in their approach, more true to their subject, more persuasive because they are more "real." This distinction depends on an important assumption—that documentarists' representations of nature are more reliable because they are true

COURTESY OF PARAMOUNT PICTURES, *STAR TREK V: THE FINAL FRONTIER* © PARAMOUNT PICTURES

To make the background for this scene (above) from *Star Trek V: The Final Frontier* as realistic as possible, filmmakers placed a fake rock wall in the parking lot of Tunnel View to stand in for El Capitan in Yosemite Valley. They then positioned the film cameras so that the real El Capitan and the fake El Capitan would blend together and in with the rest of the valley. The real El Capitan can be seen between the two actors, and is to the right of the fake wall in the bottom photo.

COURTESY OF LISA STRONG-AUFHAUSER, STRONG MOUNTAIN PRODUCTIONS, BOULDER CREEK, CA

to the historical record. Using archival and contemporary imagery, as well as human memory, they apparently more accurately convey the rich texture of a complex past. Their cameras, focused on an authentic space—not on some manufactured prop—do not lie.

Or to be more precise, the scenes they are designed to capture fall within what critic William Stott has called the "genre of actuality"; its essence is "the communication, not of imagined things, but of real things only." Yet how those "real things" are conveyed through images, dialogue, and music, and which of their viewers' emotions they tap, and thus what responses they engender, complicate our embrace of the neat dichotomies between fiction- and fact-based filmmaking. This is a point of some contention. British film producer Jerry Kuehl, for one, disputes the notion that the two forms can instruct equally well: fictionalized versions of events and people in the past deflect us from the "understanding of motives," he argues, and instead lead us "not to the historical figures [themselves], but to the writers who wrote the lines, the actors who spoke them, and the directors who orchestrated their performances." Traditional documentaries, by contrast, revolve around what he calls "truth claims," assertions that are based "on argument and evidence." They make more reliable guides to human intention and motive.[173]

Such claims for reliability, other observers argue, are themselves a kind of fiction. As film critic Philip Rosen puts it, movie-industry narratives and low-budget documentaries are linked by the very structure each employs to tell their stories. Both centralize meaning through what he describes as "internal sequenciation"—one scene builds on another to extend a particular theme or argument or narrative; each subsequent scene is legitimized by reference to the preceding ones—a methodology that structures the audience's perceptions. This narrative format indeed makes the *viewer* "a terrain to be organized," a manipulative approach that clouds a film's capacity to assert "truth claims," regardless of the kind of filmmaking involved. "All films," another critic concludes, "whether they are labeled fiction, documentary, or art…are structured articulations of the filmmaker and not authentic truthful objective records."[174]

It is no less true that audiences have been differently organized over time; the images presented within documentaries, and the reactions they have evoked, have assumed certain patterns that identify shifting modes of cultural expression. Changes in culture also have influenced how we experience documentaries shot in another era. Those from the 1930s, for instance, may not speak to us as powerfully as those crafted out of the materials of the age within which we live. I have been regularly reminded of this whenever I screen films for an undergraduate seminar in environmental history. My students, for example, are considerably less engaged by and more critical of the imagery and argument embedded within Pare Lorentz's *The Plow That Broke the Plains* (1936) than they are of *The Wilderness Idea* (1989), the film Larry Hott was shooting in Yosemite when he encountered the Star Trek crew. Explaining why this is so, why we may be more readily convinced by one set of rhetorical claims than another, depends on a recognition of the intricate interplay between audience, filmmaker, and historical context.

This interplay compels us to explore three related questions: how these films and their subjects have evolved since their origins in the 1920s; what role historians and historiography have played in the crafting of this subject matter; and how as a result we might best understand the evolving relationship between these environmental documentary productions and their widening audience.

### Take One

The evolution of environmental documentaries is shaped by narrative tension. Film scholars generally locate the origins of the genre in the contrasting work of American Robert Flaherty and British filmmaker John Grierson. *Nanook of the North* (1922), *Moana* (1926), and *Man of Aran* (1934)—Flaherty's first films—offered close observations of humans living within nature. In these features, he sought to capture the Eskimo, Samoan, and Irish struggle for survival, and recapitulate (and when necessary, *recreate*) their often vanishing cultural mores, all with an anthropological eye. He was not an anthropologist, however, and many of that profession would roundly criticize him for what he did not observe about the people and places he filmed. But his partner (and wife), Frances Flaherty, would later insist that their work was anthropologically oriented in the sense that it was focused on "discovery and revelation," an observational emphasis that Grierson picked up on in his 1926 review of Flaherty's *Moana*: "Being a visual account of events in daily life of a Polynesian youth and his family, [it] has documentary value." Film historians believe this to be the first time the word "documentary" had been used in relation to film, and its use nicely linked the two figures who are acknowledged as the first to define the character of this form of filmmaking.[175]

*The truest and most human story of the Great White Snows*

*A picture with more drama, greater thrill, and stronger action than any picture you ever saw.*

REVILLON FRÈRES
PRESENT

# NANOOK OF THE NORTH

A STORY OF LIFE AND LOVE IN THE ACTUAL ARCTIC

PRODUCED BY
ROBERT J. FLAHERTY, F.R.G.S.

Pathépicture

Filmmaker Robert Flaherty was criticized for staging events and recreating cultural mores in his groundbreaking 1922 film, *Nanook of the North*. At one point, Flaherty had "Nanook," whose real name was Allakariallak, hunt with a spear instead of the gun he normally used to show how ancient Inuits hunted.

But Grierson would also be the first to criticize his friend Flaherty's work precisely because of its reflective emphasis. "His metaphor for the contrast between Flaherty's way and his own was that Flaherty used film as a mirror while he was more interested in using it as a hammer," notes critic Jack Ellis. For Grierson, film was a tool that if properly wielded could reconstruct contemporary perceptions and behaviors. To do so, he focused on collective endeavors, in contrast to Flaherty's tales of the strivings of lone individuals or single families. Take Grierson's first film, *The Drifters* (1929), a study of the British herring fleet. It recreates a day in the life of these fishermen, from their leaving port through their storm-tossed trawling along the North Sea banks to the final auction of the catch. The significance of *The Drifters*, in Ellis's words, is its unique attempt to locate labor "within the context of economic actualities" and its portrayal of the working class "with dignity rather than as comic relief." Grierson's goal was not to make these men exotic—and therefore distinct from the viewer—but to bind together subject and audience, a goal of some political importance in a society as class-conscious and riven as the early-twentieth-century Britain.[176]

Grierson used film to expose contemporary social structures and their consequences, and believed that filmmakers must interpret and shape the world around them, just as they must edit the film itself to achieve the desired effect. This vision and methodology clashed with Flaherty's renunciation of the cutting room and his editorial assumption that the preindustrial past and its peoples had a greater intrinsic value than the then-modernizing societies in which he lived. As thoroughly as Grierson embraced the "chaotic present," as fully as Flaherty spurned its materialistic temper, their friendship yet endured. As Grierson later noted, "In the profoundest kind of way we live and prosper, each of us, by denouncing the other."[177]

The differences in the two men's aesthetic choices and political orientation have been woven throughout the subsequent development of documentary filmmaking. Grierson, for instance, had a profound impact on the "social" documentaries that flourished during the Great Depression. Among those who pursued this genre was American Pare Lorentz. Although he disliked Grierson's pedantic approach to film, Lorentz nonetheless produced *The Plow That Broke the Plains* and *The River* (1937) under the aegis of the Resettlement Administration (later folded into the Farm Security Administration). This governmental sanction brought with it social and political ends: the films were to inform and convince their audiences that the nation, through an engaged citizenry and via federal planning and regulation, could resolve some of the environmental problems that had devastated the American landscape during the 1930s; among the most pressing were the Dust Bowl and Mississippi River valley flooding. Integrating

President Roosevelt's New Deal provided the "benevolent, outside force" needed to combat widespread soil erosion in the Midwest and South in the 1930s. Tree planting, as seen here in Lafayette County, Mississippi, was a popular response to soil erosion.

clips from contemporary newsreels and feature-length films with work from some of the finest cinematographers of the day, and accentuating the films' narrative structure with musical scores from composer Virgil Thompson that were a deft blend of folk tunes and modernist rhythms, Lorentz probed the historical antecedents of the then-current dilemmas, decried contemporary inaction, and proposed governmental remedies.

In *The Plow*, the rapacious energies let loose by westward migration and a capitalist economy had stripped the land of its topsoil and its regenerative capacity; over time, the farming communities that had been dependent on the land became the victims of their own inefficiencies and greed. Only the intervention of a benevolent, outside force—Franklin D. Roosevelt's New Deal—could turn this tragedy around; salvation would come with the construction of greenbelt communities, the introduction of new methods of soil conservation, and the practice of a more conservative form of agriculture.

*The River* is similarly framed: the unchecked surge of floodwaters that routinely churned the Mississippi's vast watershed, sweeping away land

and life, was a direct consequence of human error, compounded over many decades. The destructive clearcutting of the nation's forest cover, abusive cotton farming, and industrial exploitation of the landscape unleashed the deadly torrents that undercut the ability of Americans to live along the rills, rivulets, brooks, creeks, and streams that flowed into the Father of Waters. Having destroyed the land, it was no wonder that the people themselves were impoverished. The cure? Through the ministrations of the Tennessee Valley Authority and the good work of the Farm Security Administration, floods would be brought under control, the hilly, if eroded, terrain would be restored, and the human inhabitants would be redeemed.

The link between Lorentz's portrayal of an activist and compassionate government and the Rooseveltian political agenda was unmistakable, and it was the subject of considerable contemporary controversy. The staff of the Resettlement Administration was split between those who praised the intense dynamics and political overlay of *The Plow* especially, and others—including a vocal delegation of agency workers in Texas—who objected strongly to its exaggerated depiction of the environmental disaster then engulfing the southern plains. They concluded, in a memo that listed a large number of its errors of fact and interpretation, that agency head Rexford Tugwell should not show the film in the Lone Star State, for it "would arouse...rather bitter criticism of the Resettlement Agency."[178]

That these films were, in scholar Richard MacCann's words, "propaganda for the New Deal, for specific government agencies, for a certain view of history, and a certain way of looking at public affairs" is not, in retrospect, the only set of troubling issues they raise. They also were embedded with preecological cultural values about man's relationship to nature: our dominion could be established and sanctioned through rational planning and scientific management. The easy assumption that the federal government stood as ultimate protector of the physical landscape and human environment, and the implication that this was an outgrowth of the second Roosevelt's idealism, ignored the political capital an earlier Roosevelt—Theodore—had expended to advance similar ends. Moreover, by framing the issue in this manner, the films masked just how contentious the assertion of federal sovereignty was in the 1930s, a point that is particularly significant because of the strong public response to *The Plow* and *The River*.[179]

That reaction was so potent because Lorentz's work perfectly embodied what historian William Stott calls the doctrine of the documentary.

Effective because they treated the "unimagined experience of individuals belonging to a group generally of low economic and social standing in the society (lower than the audience for whom the report is made)," these films then revealed this "experience in such a way as to try to render it vivid, 'human,' and—most often—poignant to the audience." Through moral suasion and emotion appeal, the films bound together viewer and viewed. To build on this powerful, affective connection, the Department of Agriculture established the United States Film Service to expand the range of its documentary investigation and tapped Lorentz as its first director. Congress had other ideas: it initiated hearings into "the propriety and politics of the project" and slashed the service's budget, causing the outright dismantling of the new agency. Henceforth, government-sponsored, nonfiction filmmaking would be more tightly controlled, ideologically and financially.[180]

### Take Two

A consequence of the assertion of that control was that succeeding generations of nature filmmakers, seeking to avoid overt political pressures, would pursue private funding sources. After World War II a crucial source of that capital was the new television production companies and the fledgling TV networks themselves. Their money, when combined with the vast audiences that this new medium would create, and the resultant innovative filmmaking technology that has emerged over the subsequent fifty years, led to an explosion of films and videos designed to throw wide open our window onto nature. These alterations enabled the audience to see nature in greater detail and thus in unique, more intimate ways. In turn, the newfound insights altered how Americans perceived their relationship with what they might think of as the natural world—a wild place separate and apart from human civilization. The veritable outpouring of images of wildness has done "a lot of good," writes journalist-critic Bill McKibben, by which he means political good. Whether in still or moving form, whether originating in a TV show such as *Flipper* or the cinematography of Jacques Cousteau, these images have "helped change how we see the wild." Indeed, he concludes that it is "no great exaggeration to say that dolphin-safe tuna flows directly from the barrel of a Canon, that without Kodak there'd be no Endangered Species Act."[181]

Actually, he is exaggerating, but the effect of photography on our understanding of the postwar world nonetheless has been pronounced.

The modern story that those Kodak-loaded Canons have captured is really an old tale—it draws heavily on Robert Flaherty's elegiac vision and evocative landscapes and on a Griersonian critique of the human assault on nature.

It was Flaherty who gave rise to a form of cinematic narration that included travelogues, such as Merian Cooper and Ernest Schoedsack's *Grass* (1925), a study of the animal and human migratory patterns of central Iran. Building on this tradition were Martin and Osa Johnson, who spun out a large number of what critic Alexander Wilson has tagged "photosafari" films, among them *Simba, the King of Beasts* (1928), *Wonders of the Congo* (1931), *Baboonia* (1935), the very titles of which give a sense of their exotic content, reason enough for their great appeal. As Wilson observes, the Johnsons, who had financial backing from George Eastman and the sponsorship from the American Museum of Natural History, were convinced that they were "filming 'the world as it once was,'" but in retrospect their work seems an "embarrassing amalgam of bad anthropology, natural history, and adventure—a formula that meant "box office" right up to *Raiders of the Lost Ark*.[182]

Heirs to this tradition have been the televised nature shows that emerged in the early 1950s and since have become a staple of contemporary broadcasting, whether on PBS or its many cable competitors. These programs, often underwritten by the National Geographic Society and its ilk, and entitled so as to titillate—*Wild Kingdom*, *New Wilderness*, and *Wild, Wild World of Animals*—have offered safe excursions to far-distant worlds filled with strange flora and odd fauna. These exotic landscapes have been translated to Western eyes and ears by men of science (they were not necessarily scientists) who, like Jacques Cousteau, Marlin Perkins, Lorne Greene, or David Attenborough, have helped us comprehend the incomprehensible.

But it is how we are instructed to comprehend these outlandish places or peculiar behaviors we witness from afar that is so crucial to the evolution of environmental documentaries. Consciously or otherwise, many of these films have depended heavily upon a perception of a humanity that is directly at odds with nature and is in sharp conflict with its design. The relevant filmmakers may not explicitly cite Aldo Leopold, John Muir, or Henry David Thoreau, but they have absorbed these writers' conceptions of a contested relationship, and in doing so have used environmentalists' ideas to organize and frame their film narratives. So as Cousteau dived into tropical waters to reveal the abundance of marine life in sun-kissed seas that otherwise we would never have seen, as intrepid *National Geographic* photographers have snapped shots of primitive people or primeval landscapes much removed from our living rooms, always there was the presumption that these places and societies—at once both like us and not—were not long for this world. The moment civilization descended on these pristine environments and naive human communities, they would be forever lost, a point the very presence of the looming film crews foreshadowed: the whir of the video camcorder announced that the end was at hand.

Such worries, Alexander Wilson argues, were perhaps first inserted in the nature films that Walt Disney produced and in his studio's educational series, *Filming Nature's Mysteries* (1956). It was the Bard of Anaheim who pressed nature's case with "new urgency," who believed it was necessary to "'get wildlife on film before civilization could wipe them out.'"[183]

His concern was perfectly attuned to a major alteration in mid-century American culture. Beginning with the Depression, there was a massive surge of population out of rural, agrarian cultures and into urban, industrialized economies. Poor whites and African Americans moved from Appalachian valleys and river deltas north to Detroit and Chicago; sharecroppers victimized in the Dust Bowl, who once had been the subjects of Pare Lorenz's film studies of the Plains states, fled to the Pacific Coast. This pattern of displacement and migration widened during the economic boom of the war years, so much so that this is when Texas, long the locus of the legendary, frontier-roaming American cowboy, officially became an urban state.

What happened, however, when ranch hands became factory workers or aircraft assemblers, when they dug not fence posts but oil wells? The impact of these significant alterations in daily life were particularly manifest in the changes in the relationship between humans and the animals with whom they once had shared a more rural environment. Wilson and other scholars contend that urbanization over the past two centuries, which certainly intensified in the United States following World War II, led to the excision of animals "from the everyday lives of most Westerners, an excision recorded in the subsequent proliferation of zoos and animal toys and animal movies." We have created these institutions and material representations to maintain a record of "lost species" and simultaneously, to reintroduce "the idea of nature" into our lives. Reflected in this effort is a poignant human urge: wildlife films, Wilson observes, "reveal a deep desire simply to be in the world, a world "beyond our skins if not beyond our culture." This quest has become all the more complicated as we have become an even more intensely urbanized people, leading, ironically, to a further proliferation of this genre of film.[184]

### Take Three

The cultural significance of environmental documentaries is compounded by a crucial shift in how contemporary Americans obtain information about the world around them. As film historian Alan Rosenthal notes about documentaries in general, they have "become a major—possibly the most important—means for learning about the past. In an age when reading is in decline, the documentary—much more than the theater, newspapers, or feature films—may well be the only serious access people have to history once they have left school." This trend has accelerated with the "growth of cable and satellite-based television since the mid-1980s," historian Robert Brent Toplin has confirmed. These new forms of telecommunication have "created a multichannel universe that permitted greater attention to audience niches. A proliferation of channels enabled TV producers to direct attention to much smaller clusters of audience interests than NBC, ABC, and CBS could handle in the days of their shared monopoly over commercial entertainment." Environmental filmmakers are among those documentarists who have taken good advantage of these developments and whose work has appeared regularly on the History Channel, A&E, and on PBS. That being so, it becomes vitally important to think through how these "visual" narrations are constructed for these new audiences, as well as who and what lies behind their construction.[185]

That set of questions drove a heated exchange between historian Donald Watt and film producer Jerry Kuehl about who is the most appropriate author of a film and who controls its central narrative thrust. Watt believes it is "self evident that the making of a nonfiction film or television program on a historical theme is as much an exercise in historiography as is the composition of a learned monograph…" But it is no less obvious, he remarks, that a "historical statement made audiovisually is different from one made in writing. The tempo is different, there can be no recall, no flipping of the page, no elaboration of parallel themes by footnotes or parentheses." These profound distinctions are then magnified if the filmmakers and historians are insensitive to each other's professional foci. At its worst, their relationship can devolve into a bitter dispute over film content and audience expectations, the two being parts of a whole; as Watt acerbically notes, "the biggest problem" has to do with the "state of mind of those who direct the media, who cannot believe that waiting in front of their sets is an educated, interested, mass audience...." Producers dumb down historical content to appeal to the lowest common denominator, a tactic that undercuts historians' contributions. The best that can be hoped for is to develop a relationship that is more of a partnership, one that shades into "symbiosis, where each understands, even if he cannot practice, the craft of the other," making "audiovisual historiography" a kind of "bimedial art, like ballet and opera."[186]

Such a harmonic convergence is rare. Only at England's chronically underfunded Open University has it been "most closely approximated," Watt observes; there, budgets allowing, film producers collaborate with historians, who write the script and select "the material they wish to see incorporated into the film." The pronoun "they" is revealing of a shift in the control of production. Historians, in Watt's scenario, are the creative forces driving a documentary's development from idea to film; their assumptions dominate the text's construction and the context in which the final product should be viewed (and interpreted).[187]

It is precisely Watt's claim of historians' preeminence that Kuehl believes has so troubled relations between "academic historians and producers of television documentaries" and made their collaboration so "uneasy." Although he acknowledges that misapprehension of each other's roles is usually what undercuts joint ventures, he is not ready to relinquish to historians the control that Watt believes is their due. He resists because he does not believe scholars understand the special requirements of television filmmaking. Consider that most "integral part of every documentary," the commentary: "who should write it, how should it relate to the film, to whom should it be addressed, and above all, what should it contain?" The answer to these critical queries is shaped not by a historian's professional competence but by the amount of time usually allotted to this form of documentary—most have but fifty minutes to tell their tale. That being the case, he estimates that "a commentary of between one thousand and fifteen hundred words is quite long enough—any more, and the film...will appear to viewers as dense, over stuffed. They will be repelled, not informed," an observation Kuehl hopes is "quite sobering to an academician." Film writers only have enough time to give the equivalent of a "*fifteen-minute* lecture," and that lecture must conform to this reality:

"because of their brevity, they cannot be in any real sense exhaustive or comprehensive." More evocative than coherent, the commentary is neither an "independent literary form" nor a means by which to narrate "complicated narrative histories." It is, moreover, "quite hopeless at portraying abstract ideas." For this work, historians need not apply.[188]

What professionals of the past should do instead, he concludes, is to stick with the academic terrain they know best: rather than "trying to replace mass television history of our day with their own mandarin versions, they should concentrate on doing their jobs as historians as well as they can, so that the history they write will be as good as it can be, so that popular accounts which we can provide will be as true as they can be." Note Kuehl's pronoun of choice—"we"—which signals, as if the rest of his conclusion did not, that documentaries are the domain of filmmakers, who better understand how to engage mass audiences.[189]

Another who identifies this point as the source of creative conflict is Richard White, an environmental historian who has "worked as an adviser or been a talking head for a half-dozen or more documentary film projects." He believes there "are always moments when the historian confronts the limits of television, filmmakers, and the television audience." Their limitations are many: the documentarist's demand for "iconic" stories that streamline the film's narrative but forsake historical complexity; the producer's fondness for a "binary cast of characters" whose conflict moves the storyline along and thereby maintains audience attention; and the "cavalier" use of evocative photographs, however out of context their use might be. Filmmakers "want pictures that will carry the idea," he asserts, and given "the choice between a somewhat deceptive photograph and a talking head, the photograph wins almost every time."[190]

My relatively limited experience working on several environmental documentaries confirms many of White's insights. That said, the collaboration between filmmakers and historians is not always antagonistic. On three projects for Florentine Films—*The Wilderness Idea: John Muir, Gifford Pinchot and the First Great Battle for Wilderness* (1989), *Wild by Law* (1991), and *Divided Highways* (1997)—I joined a host of scholars to offer insights into the struggle over Hetch Hetchy, the intellectual and political antecedents to the Wilderness Act of 1964, and the profound impact of the interstate highway system on the built and natural landscapes of the United States. We were extensively interviewed off-camera on the relevant topics, and our comments served as one basis for a lengthy "script" whose various iterations we then read and critiqued. This was a fascinating perspective from which to observe the narrative unfold, watch the voice-over commentary be refined, and track shifts in the placement of some of the set images. Many of these initial efforts would be modified, more or less extensively, once the interviews took place before the cameras. But it was evident that the questions the producers asked us during the taping also had been defined in good part by historical research and our subsequent conversations with the filmmakers. The resultant films were a fruitful outcome of this intense interaction between filmmakers and historians, suggesting that it is possible to construct environmental documentaries that engage and educate television audiences and remain faithful to each field.[191]

Yet it was also clear that in these instances the control of the final product lay very much in the hands of the films' producers. Proof of this lies in White's quip about the ability of a photograph to trump a talking head, and in my ready resort, after viewing the films in which I (briefly) appeared, to this Hollywoodian wail—they left my best lines on the cutting room floor! That there is a cutting room is a critical reminder of how essentially different modern documentaries are from those produced by Robert Flaherty: he eschewed editing (though he was not above doctoring scripts and coaching his subjects). That modern documentaries are thoroughly, self-consciously edited reminds us that their viewers must take care when watching them, especially those as emotionally charged and politically driven as environmental documentaries can be. We have to learn to read films with all the subtlety we apply to literary analysis.

In practical terms this requires historians—public and otherwise—to assess which archival resources were consulted, which experts were interviewed, which eyewitnesses spoke on screen, and how those interviews were framed. Knowing this helps identify the film's larger argument and organizing structure; for the same reason it is important to clarify who sponsored and paid for the film.

Gaining clarity on a documentary's point of view is just as crucial. It "must tell a story that will hold the viewer's interest," argues historian James G. Lewis, and "it must do so in a clear, concise manner so that the subject matter is accessible to as many people as possible." Because accessibility determines content, it is all the more important that the intended audience recognizes the formulaic quality of the documentaries it watches on PBS or cable; only by doing so can viewers develop a more detached examination of a film's interpretative agenda.[192]

One of the hardest elements to detach from is the omniscient narrator—he or she is the guide who unifies the audience's reading of a film. It is for this reason that the use of such a voice has become so controversial. Many critics decry the insertion of a consistent, authoritative narrator, believing that this "voice of God" serves as a divine intermediary between subject and audience; a "didactic reductionism" dominates the viewers' aural experience. Surely this situation is compounded when a celebrity stands in for a historical character. In *Rachel Carson's Silent Spring* (1993), Meryl Streep speaks for the embattled author, a choice that confounds— are we persuaded of Carson's heroics because of her incisive mind and enduring legacy or because her accomplishments (and identity) are now intertwined on film with the vocal persona of one of our era's greatest actresses?[193]

A different form of spin revolves around the selection of background music to emphasize mood, sensibility, and moral standing. In *The Wilderness Idea*, which pits the Hetch Hetchy dam's chief opponent, John Muir, against one of its proponents, Gifford Pinchot, the musical score becomes a box score. For Muir the key is minor, a tragic tone that befits his valiant, if failed, attempt to thwart the drowning of his beloved valley. Whenever the triumphant Pinchot appeared in the film, in stark contrast, he is enveloped in a giddy, waltzlike air, as insubstantial a piece of music as his victory apparently was unconscionable.

Establishing such critical distance, and the knowledge on which its development depends, should not be a pursuit only for academicians. Environmental and public historians have an obligation to carry their insights into the civic arena because it is there that national conversations about the meaning of contested pasts are debated. For instance, a prototypical example of public historians' "teaching" film in public venues to ensure a more critical understanding occurred with the screening of *From Rosie to Roosevelt*. In public libraries across the country, scholars showed the film about the American experience during World War II and then led discussions exploring the relationship between visual and written evidence. On a smaller scale, but with similar intent, those involved in the making of *The Wilderness Idea*—including producer Larry Hott and many of the consulting historians—staged public showings of the film. I participated in four such evenings, in Massachusetts, Oregon, Washington, and Texas, and each time when the house lights came back on, discussion erupted. In the Pacific Northwest, then at the height of the spotted owl controversy, the film spoke vividly to citizens at logger-heads over the impact of intense harvesting of public forested lands. There, but also in other regions not buffeted by such contentious environmental issues, debate swirled around how that film shaped its arguments. Even those who applauded the Muirite vision that informed *The Wilderness Idea* could be critical of its intellectual bias.[194]

Those experiences have taught me a valuable lesson: being aware of a film's many manipulative devices only increases its heuristic value. That seems paradoxical, but only if we assume that "truth" is the single best teacher. Blatant falsehoods, historical inaccuracies, overdramatized reen-actments, subtle intercuts: all are vehicles by which to explain authorial intent. How producers argue their points tells us a great deal about their interpretation of the relationship between humanity and the landscape in which we are enmeshed, and that knowledge permits a more sustained analysis of the central environmental dilemmas that lie before us.

CHAPTER THIRTEEN

❧

# Trees, Sprawl, and Urban Politics*

Trees grow in context. Their individual and collective lives are determined by a welter of physical factors—among them, water and humidity, soil, topography, climate, altitude, and ecological niche. But their capacity to grow is also defined by the political landscape, a man-made terrain of ideas, ambitions, and symbols. What is a tree? What we *see* in one.

Trying to define that human dimension is what urban forester E. Gregory McPherson was getting at when he argued that "Trees have always been viewed as ornamental, but they are now seen as providing social, economic, and environmental benefits as well." This widening perspective among foresters is largely a consequence of explosive rates of urbanization in the United States since World War II and the staggering increases in "air pollution, more congestion on highways, loss of biodiversity, and occasional shortages of energy, water, and other resources." As they have moved out to the urban fringe, Americans have carried the city with them, further fragmenting much-needed forested habitat and tree canopy. These environmental challenges will only intensify. "During the past century we have learned how to manage forests for spotted owls and songbirds," McPherson confirms. "Yet we have not succeeded in protecting green space near cities or creating environments that make people happy."[195]

San Antonio, Texas, the nation's eighth-largest city, is among those that have failed to maintain or build a sustainable urban landscape. Home to more than 1.3 million people (and as many cars), set within a county of 1,248 square miles, it is the archetype of contemporary sprawl. Surrounded by concentric rings of freeways, its concrete corridors are lined with big-box shopping malls, restaurants, hotels, and gas stations; fly ramps arc across the sky, and drainage culverts curve beneath the elevated

---

* This is a substantially revised version of a talk, "Changes in the Landscape," copresented with Douglas Lipscomb at the National Urban Forestry conference, San Antonio, Texas, 2003.

infrastructure, a forbidding territory that only the automobile can navigate. The impact of this development on the region's ecological health is registered in the sharp decline in tree canopy. A 2002 American Forests aerial analysis of metropolitan San Antonio reveals that between 1985 and 2002, acreage of full tree canopy plunged from 201,000 to 156,000, "a loss of over 22% of the densely forested areas." The amount of medium-density canopy was nearly halved. The obvious conclusion: "as new development occurs, tree canopy is not being preserved."[196]

Those data confirmed what local landscape architects, urban foresters, and environmentalists had long suspected and been trying to counter since the late 1980s through calls for a strong tree preservation ordinance and a rigorous set of smart-growth regulations. Their campaign was energized in 1993 when bulldozers, preparing the site for a Kmart megastore, leveled groves of old-growth live oak. The toppled trees prompted City Councilman Howard Peak to blast the company's actions as yet "another example of irresponsible development." Fellow Councilman Bill Thornton cast the episode in tragic terms: "These were 200-year-old oaks, and they may replace them with 3-inch saplings…but it's just not the same." Decrying Kmart's "expression of contempt for the residents of San Antonio" was *Express-News* columnist Mike Greenberg. His demand that the city enact an ordinance to protect and preserve trees was seconded by a host of environmental groups in the aftermath of what one critic dubbed the "Texas Chain Saw Tree Massacre."[197]

The outcry had only a modest impact. When in the mid-1990s the city council passed a tree preservation ordinance, its provisions were so minimal that developers and environmentalists concurred that its requirements easily would be met. To secure the ordinance's passage, a divided council finally had to agree that the law be reevaluated and strengthened in five years. That deadline was not met, and because the council delayed revising the ordinance until spring 2003, much of the land it was to regulate had already been permitted for construction and thus was not governed by its regulations. By then, the Kmart superstore, which a decade before had sparked the initial demand for a preservation ordinance, had closed.[198]

Those recent controversies are part of a three-century-long interaction between the natural and built environments in San Antonio. Recognizing that there is such a lengthy chronology is critical because urban forested landscapes in general are "social creations and…respond

Developments like this one north of San Antonio have left large areas of the city deforested. The suburban sprawl has compromised the Edwards Aquifer's recharge capacity and generated pollution that has threatened water quality.

to social patterns of power and influence." An analysis of the complex and extended interplay between politics and ecosystems in San Antonio can clarify why the city has had such a difficult time enacting smart-growth ordinances and promoting a healthier environment.[199]

## Spanish Urbanism

Frederick Law Olmsted observed the local intersection of culture and environment when he rode south into San Antonio in 1857, a ride that had offered little vertical relief: "The trees were live oaks and even these were rare....The ground-swells were long, and so equal in height and similar in form, as to bring to mind a tedious sea voyage, where you go plodding on." But then he topped a rise and "saw the domes and white clustered dwellings of San Antonio below us. We stopped and gazed long on the sunny scene." As he entered the town of four thousand, he was struck by the contrast between its barren exterior and lively core. The built environment was a jumble of styles and influences that mirrored the babble of tongues Olmsted heard in the crowded streets, packed plazas, and boisterous alleys; the citizens seemed in perpetual promenade.

Recognizing that this vivid tableau, courtesy of Spanish urban planners, was unusual, he likened San Antonio to another former Spanish town, New Orleans, in its picturesque feel and antiquated foreignness.[200]

Certainly the community's eighteenth-century planners had developed a spatial structure that compelled citizens to live, work, play, and pray within a civic center. Devised according to the urban form prescribed in the Spanish Empire's code, the Law of the Indes (1583), the local *Plano de la Población* commenced with the city's San Fernando cathedral, sited just west of the San Antonio River. From its front door, the *calles* and plaza were marked off, as were surrounding public buildings and individual housing, all locked in a westward-facing grid.[201]

But there was a flaw. "The land to the west of the presidio, the location of the said villa, has no facilities for irrigation," Captain Juan Pérez de Almazán wrote, and because a crop failure would have imperiled the colonial outpost, he spun the initial plan on its axis, so that the church's front door faced east. This reorientation pushed the villa into land situated between San Pedro Creek and the Big Bend of the San Antonio River, a revision that had disastrous consequences: it placed the town in the floodplain. In 1819, a century-level flood ripped through the village, killing many people and animals and destroying property. Yet for all its force, that flood, and the others that followed, did not alter the walking city's character, in large part because San Antonio grew so slowly. An estimated twenty-five hundred people had called it home in 1803, and only twelve thousand did in 1870. It easily absorbed newcomers by expanding the original Spanish design, wherein residential areas and commercial activity fronted shared open space.[202]

Almazán's cityscape could not withstand the railroad, however, which arrived in February 1877 and announced a new urban order. Within a year, mule-drawn trolley cars were hauling passengers between the railroad station, downtown offices, and new, more-distant neighborhoods. San Antonio's population grew quickly, rising to more than fifty thousand residents in 1900, and nearly a hundred thousand a decade later. This boom established a new white majority, many of whom moved to streetcar suburbs occupying high ground to the north and east of downtown. This increased distance between work and home also intensified social prejudice and class distinctions. In a walking city, the rich and poor were compelled to live in close quarters; that proximity was no longer necessary in the Age of the Iron Horse.[203]

By the 1920s, the automobile had pushed the development of exclusive suburbs beyond the city limits. The Woodlawn and Beacon Hill neighborhoods to the northwest of downtown were constructed at the same time as Terrell Hills and Olmos Park, lying north and east; in between these latter two communities was an older streetcar suburb, Alamo Heights. Twenty years later, the city began to annex some outlying neighborhoods, but because it also funneled funds into expressway construction, it accelerated northern sprawl first to Loop 410 and then to (and beyond) Loop 1604. Predictably, this exurban migration spawned the same social problems and environmental stressors those fleeing the city had hoped to avoid—more roads, more cars, higher crime, increased pollution, fewer trees.

That rapid dispersal, as in other postwar American cities, had a consequence: it imperiled regional water supplies, for beneath the city's northern, tree-studded hills lies the recharge zone for the Edwards Aquifer, essentially the community's sole source of water. Because sprawl compromised the aquifer's recharge capacity and generated pollution that threatened water quality, citizens fought against these threats with a series of public policy initiatives.[204]

In the end, none of the challenges proved particularly successful, although not for any lack of sustained effort on the part of those concerned about how to create a greener city. In 1991, in response to the manifold environmental and spatial problems resulting from San Antonio's postwar development, the city council established a committee to develop a new master plan, and six years later it formally adopted a complex policy that laid out several related goals: the sustainability and conservation of natural resources, including the preservation of trees and tree canopy; the management of growth to ensure a balance between the community's geographic regions as well as between inner-city revitalization and new peripheral development; the preservation and enhanced livability of neighborhoods; and the establishment of urban design standards, such as the development of a citywide park system and the fostering of alternative transportation systems. The desired outcome was a city containing a more vibrant center and denser development on the periphery.[205]

The Master Plan Policy Advisory Committee was subsequently formed to identify impediments to implementing the new policy that would result from conflicts with the City of San Antonio's 1987 Unified Development Code. Not surprisingly, it identified many hindrances. The

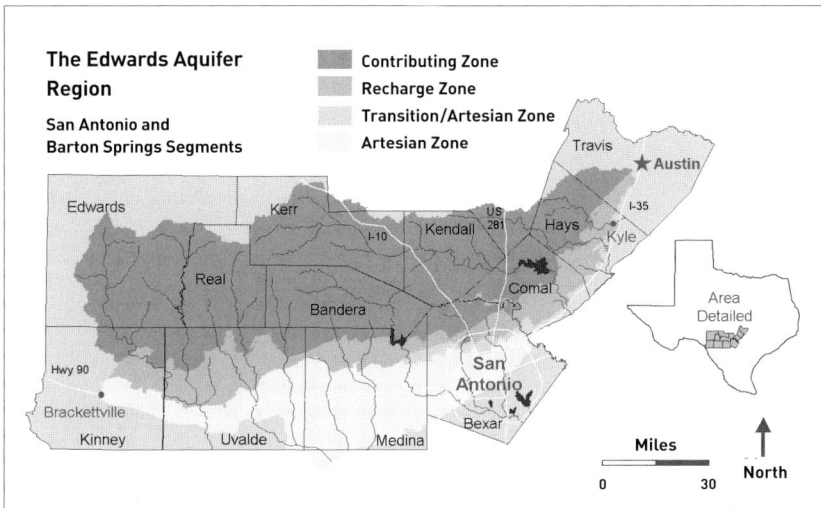

**The Edwards Aquifer Region**

San Antonio and Barton Springs Segments

Contributing Zone
Recharge Zone
Transition/Artesian Zone
Artesian Zone

Edwards
Kerr
US 281
Kendall
Hays
I-35
Travis
★ Austin
Kyle
I-10
Real
Comal
Bandera
Area Detailed
Hwy 90
San Antonio
Brackettville
Bexar
Kinney
Uvalde
Medina

Miles
0          30

North

A map of the Edwards Aquifer, showing the recharge and contributing zones. The aquifer supplies a combination of agricultural and ranch land and areas of dense population—parts of ten counties in all. The city of San Antonio, which has a population of more than 1.3 million, derives its entire water supply from the Edwards Aquifer.

1987 code, for instance, had so segregated land uses and separated related activities such as living, working, shopping, and recreation that, in the absence of viable alternative mass transportation, it effectively mandated the use of a car. This automotive dependency produced acres of new asphalt parking lots downtown and generated the grading out of the hilly, tree-filled topography along the city's northern rim. Sanctioned under the provisions of this earlier code, for instance, had been the 1993 bulldozing of oak trees on the Kmart building site.[206]

Reinforcing the automobile's primacy were a host of related development standards, such as excessive minimum parking requirements, building setbacks, street widths, and wide-turning-radii requirements. Promoting a pedestrian-hostile environment, the 1987 code's development standards also increased the loss of natural habitat to surface paving, with corresponding threats to water and air quality from parking lot runoff and smog from the increased vehicle miles driven.

So at odds was the 1987 code with the goals and policies of the 1997 master plan that the advisory board concluded that crafting an entirely new code would be the only way to enhance environmental protection and construct a more sustainable, livable community. Especially supportive of this conclusion was Councilman Howard Peak, a planner-turned-politician who became mayor in June 1997. Under his leadership, the city council in September 1999 hired Freilich, Leitner & Carlisle (FL&C), a Kansas City law and planning firm, to prepare a draft revision of the 1987 code. Between October 1999 and May 2001, stakeholder groups met with Mark White, an FL&C consultant, in a series of often-raucous and marathon public meetings and helped review innumerable proposed drafts of the code. The scope of the subsequent revisions, which the city council adopted in May 2001, was substantial; they established a series of improved design standards for conventional development that inverted some of the previous code's principles:[207]

- maximum building setbacks and mandatory planting of street trees to encourage the development of pedestrian-friendly streets;
- limits on parking spaces to conserve more natural habitat by creating more compact or denser development patterns; and
- park and open-space requirements to conserve meaningful amounts of open space in a city that has one of the nation's lowest park acreage per capita; the goal is to provide parklands in residential zoning districts at the rate of one acre per 114 dwelling units.[208]

The Unified Development Code of 2001 adopted some of the principles of new urbanism and its corollary, "smart growth." Founded in the early 1990s, the Congress of New Urbanism established a set of urban design principles that, taken together, encouraged "the restoration of existing urban centers and towns within coherent metropolitan regions, the reconfiguration of sprawling suburbs into communities of real neighborhoods and diverse districts, the conservation of natural environments, and the preservation of our built legacy."[209] In San Antonio, these general ideas found specific expression in a set of new development options that included "use patterns" and new zoning districts to counter the city's postwar sprawling development:

- *Traditional neighborhood development*, which mandates a mix of uses in a compact pedestrian-oriented streetscape and is intended to reduce reliance on automobiles, decrease the number of parking lots, and moderate their combined environmental impact.
- *The conservation subdivision*, which establishes overall project densities rather than minimum lot sizes, allowing developers flexibility in the configuration of the development; it mandates the preservation of

natural resources through minimum open-space requirements, such that fifty percent of the overall tract is dedicated to conservation land.

- *The commercial retrofit*, which aids the redevelopment of existing shopping centers, big-box retail sites, and other sites characterized by large expanses of surface parking, into a new development pattern that is pedestrian friendly and allows the development of existing parking lots, limits the number of parking spaces, and provides a density bonus to encourage such activity.

- *The infill development zone*, which focuses on the development of bypassed parcels, allowing for development without extending the city's limits into surrounding undeveloped areas.

- *The transit-oriented development zone*, which encourages a mixture of residential, commercial, and employment opportunities along transit corridors; it also regulates uses and development to create a more intense, built-up environment that is oriented to pedestrians and encourages mass transit; its goal is to reduce car trips and their effect upon the environment.

For all its comprehensiveness in scope and consistency of purpose, the revised code has had no noticeable impact on developmental patterns in San Antonio. It may never have much sway, either—a consequence of political realities that from the outset have compromised its regulatory authority. The state's generous grandfathering laws, for example, will delay for years the implementation of the more stringent development standards. Those standards have been further eroded because of the clout of local developers who persuaded the city council to scuttle important regulations before they were imposed; the maximum building setback, for instance, never was enacted. The hope that the new code would enhance the quantity and quality of urban parks was been scuttled, and "options" to regulations gutted many of the smart-growth elements. Once mandatory, new "use patterns" are now voluntary; gone, too, are the incentives that had been built into the code to stimulate better design of streets and neighborhoods. A critical component of new-urbanist planning—mixed-use projects—vanished in the face of opposition from the developers, contractors, and financial institutions. To make clear their desire to undercut the city's capacity to enforce any remaining provisions of the new code, in June 2004 developers filed two dozen joint lawsuits alleging that the city's planning codes, some of which date back to 1924, are illegal.[210]

COURTESY OF JON THOMPSON

San Antonio grassroots activists used this postcard to prod the local media into examining the cozy relationship between developers and the city's political leaders. The postcard claimed the image showed the "Obsceno Ridge Subdivision: a project of PullTree Homes," a play on the name of a home construction company.

Trimmed back, too, was the 2003 revision of the city's Tree Preservation Ordinance. Its new protections and regulations had taken more than two years to negotiate, and Mark White, the FL&P consultant who had devised the city's new code, brokered the new document through five grueling public hearings and city council discussions. As written, it was stronger than its predecessor, establishing criteria for the preservation of "heritage trees," granting greater protection to forest understory, establishing wider root-protection zones and more pervious ground cover, and increasing the number and percentage of trees to be saved on commercial sites and housing developments.[211]

Those enhanced requirements came with loopholes and "options" that have weakened its regulatory impact. One such option allowed developers to make a payment to the city in lieu of protecting tree canopy. The first major project to employ this option was the Toyota Motor Company, which in January 2004 broke ground for a massive Tundra truck plant on a 2,654-acre site on the city's south side. So that it would not have to preserve or mitigate one hundred percent of the 1,260 acres of tree stand on its property, Toyota opted not to conduct the rigorous tree survey that the ordinance mandated, opting instead to utilize an alternative "tree-stand delineation method" to determine which trees it would protect. According to Assistant City Manager Chris Brady, this loophole "requires that [only] 25 percent of the 1,260 acres be preserved," for a total of 447 acres of trees and understory. The agreement between the city and Toyota led Roddy Stinson, an *Express-News* columnist, to predict that a new precedent had been set—other "developers of all large tracts of land will be allowed to choose the tree-stand delineation 'alternative'"—and that this would have stark results. San Antonio's dramatic canopy loss, captured in the American Forests' 2002 aerial survey, would continue apace.[212]

## Future Prospects

The very urbanizing pressures that played such havoc with San Antonio's efforts at city planning and urban forestry may produce some good. "As Americans become increasingly urban, urban forests become increasingly important," forester E. Gregory McPherson has argued. To secure the many benefits of these forests and attendant green space—decreased storm runoff, improved air quality, reduced energy consumption, enhanced aesthetic pleasure—will require a renewed commitment to stewardship, a desire to connect people "to nature and to each other." And if this new urban land ethic "is going to emerge during the 21st century," he predicts, "it will spring from our cities."[213]

Perhaps even in San Antonio. Evident in the years of its contentious public meetings over two tree preservation ordinances, and the controversies swirling around the enactment of a more progressive unified development code, has been an elevated consciousness among some San Antonians of the environmental troubles building in the city's central core and on its expanding fringe. For instance, the Alamo Forest Partnership, founded in response to the 2002 American Forests tree survey, has committed itself to increasing canopy to thirty-five percent. If "we are careful about what trees we take out and combine that with a tree-planting program," said Calvin Finch, a member of the partnership's steering committee, "we should be able to reverse the trend of losing tree cover." In this stated ambition to protect trees, in the broader fight to secure more parklands, and in the hope for the development of smarter and more sustainable growth patterns lies an enduring belief that human beings are active agents in the creation of healthy communities.[214]

That ethical impulse is woven into Aldo Leopold's assertion that "conservation is the state of harmony between men and land." Easier to speak of the need for an equilibrium, the famed conservationist recognized, than to achieve. And it would be gained, and held, only when we "cast off…the belief that economics determines *all* land use," he asserted. "An innumerable host of actions and attitudes, comprising perhaps the bulk of all land relations, is determined by the land-users' tastes and predilections, rather than by his purse." But developing the political legitimacy for and cultural sanction of Leopold's perspective has been tremendously difficult, as San Antonio's environmental history suggests. Just how tough it has been is reflected in the double negative Calvin Finch employed to shape his vision of San Antonio's future: "We don't want a city with no trees."[215]

AFTERWORD

❧

# *An Open Field*

Perhaps only a poet could capture the bewildering shifts in the climate of south Texas. Sidney Lanier, convalescing there in 1873, sketched out a tale about "those remarkable meteorological phenomena called 'northers'" that sweep across the region with peculiar fury. Imagine "riding along the undulating plains around San Antonio on a splendid day in April," he wrote in an evocative essay first published in *Southern Magazine*, for that is "when the flowers, the birds, and the sunshine seem to be playing a wild game of which can be maddest with delight, and the tender spring-sky looks on like a mother laughing at the antics of her darlings." This tranquil moment seemed to stretch on, even as the day heated up: "presently you observe that it is very warm. An hour later you cannot endure your coat; you throw it off and hang it around your saddle," but this brings scant relief. Soon "the heat is stifling, thermometer at ninety degrees, which on the windless prairie with the Gulf moisture in the air, is greatly relaxing," especially for consumptives like Lanier. Then, while standing "on an elevation in the hope of getting some breath of air, suddenly you observe a bluish haze in the north," and within a "few moments a great roar advances…and presently the wind strikes you, blows your moist garment against your skin with a mortal chill." Illness lurked for any who did not quickly mount up and "make for a house as fast as your horse can carry you," or failing that, seek shelter in "some thicket of mesquite in a ravine under the lee of a hill, for within an hour the temperature may plummet 40 to 50 degrees." There would be a calm after the storm: in its energetic wake, the tempest would spin off succeeding days of bracing temperatures and crystalline skies, a fair return for a thermometer that "cuts such capers."[216]

It should have been during a similarly punishing and mercurial moment in late-twentieth-century San Antonio, enveloped in a "furious storm of rain, of hail, or of snow," that I initially encountered Richard White's seminal essay, "American Environmental History: The Development of a New

Historical Field." Such a convergence of art, life, and weather pattern might have defied reality, but it would have made for a fabulous narrative opening. That said, like the norther's rush, his article blew me away.[217]

Trained as an intellectual and cultural historian, and backing into environmental studies as a result of research into the life and activism of Gifford Pinchot, I read White's careful dissection of the emerging field and discovered (not for the last time) just how much I did not know. The contours of this new academic landscape were as unfamiliar as its language was confusing and its signifiers obscure; like a naive Lanier, who had trotted out onto the plains thinking one spring day was just like another, I did not have a clue where I was or what I had come upon.

This disorientation was also alluring. As it probed the broad range of topics that made up environmental history in the mid-1980s, from the biographical to the urban, from the ecological and aesthetic to the hydrological and the wild—only some of which I knew in passing—White's survey proved a beguiling guide for the perplexed. Most immediately understandable (and thus comforting, I suppose) was its deft sketch of the discipline's intellectual evolution, beginning with the varied influence of writers as diverse as Frederick Jackson Turner, Walter Prescott Webb, and James Malin on those in the 1960s who were developing the modern field of environmental history. This new stage of scholarship, a result of the pioneering work of historians Samuel P. Hays, Roderick Nash, and Donald Worster, among others, had coincided with and had drawn not a little strength from the emergence of the environmental movement. Here, White introduced a theme central to his article (and, as it has turned out, to the discipline's transformation): the powerful and compelling connection between scholarly analyses of the environment and a force then gaining strength in American politics—the environmentalist ethos. In this academic field the grass was greener; it had a profound dose of social relevance, no small consequence to those who were mapping this new terrain, or who, like me, had simply stumbled upon it.

But although White, then at the University of Utah, acknowledged the excitement and urgency that advocacy could bring to once-stolid academic agendas, he also understood the associated costs. When historians conflated the past and present, when they probed other times in search of representative arguments for which they felt considerable affinity, they perforce blurred fundamental distinctions between historical moments and missed the shifts in context that defined past environments and the human conceptions of them. "The past may be another country," he cautioned, "but for some authors a transcendent nature can wash away the boundaries that time creates" so as to locate "a universal language shared by author and subject."[218]

About wilderness has this been especially true. The idea of it "has become the highway to the American psyche most favored by intellectual historians," White observed. "Whether hated and feared, loved, or best of all, beheld with tortured ambivalence, wilderness has become the mythic core of the American experience." So it remains, though perhaps it is more accurate to say that our response to wilderness has intensified in the intervening years. This intensification has something to do with the apparent diminution of acreage defined as wilderness, the shrinking of intellectual horizons, and the hardening of political rhetoric. Now ever more fraught with cultural meaning, more embedded with psychological value, more infused with spiritual sensibility, wilderness has become the national raw nerve.

Raw enough, in any event, that those who have probed our manifold connections to wilderness, or who have wondered about its alleged sanctity, have come in for a touch of partisan mugging. So William Cronon discovered (and surely anticipated) when he delivered a talk entitled "The Trouble with Wilderness: Or Getting Back to the Wrong Nature" at the 1995 meeting of the American Society of Environmental History; his essay subsequently was published in a variety of venues, most provocatively as the centerpiece of a forum in *Environmental History*.[219]

Fascinated by the American impulse to deify wilderness, Cronon argued that this long-standing sanctification—with its roots reaching deep into nineteenth-century Romanticism—reinforced an artificial and politically dangerous distinction between the natural and the human. "Idealizing a distant wilderness too often means not idealizing the environment in which we actually live, the landscape that for better and worse we call home." Rather than continuing to embrace what he called a "set of bipolar moral scales in which the human and the nonhuman, the unnatural and natural, the fallen and unfallen, serve as our conceptual map for understanding and valuing the world," he proposed we accept instead the "full continuum of a natural landscape that is also cultural, in which the city, the suburb, the pastoral, and the wild each has its proper place…." Only through such an inclusive perception would we be able to "recollect the nature, the culture, and the history to make the world as we know it," a knitting together that would have important policy implications. If wilderness "can start being as humane as it is natural, then perhaps we can get on with the unending struggle to live rightly in the world—not just in the garden, not just in the wilderness, but in the home that encompasses them both."[220]

Cronon's considered search for a resolution to what White sardonically dubbed Americans' "tortured ambivalence" about wildness provoked decidedly unambivalent responses (thereby underscoring White's original insight about our tormented state). Many environmental historians—including Samuel Hays and Michael P. Cohen, who rebutted Cronon in the *Environmental History* forum—took him to task for a series of sins, real and imagined. Some angrily dismissed his analysis of the cult of wilderness, suggesting that he had confused the physical place with its mental representation; others took offense at the polemical nature of his argument even as they questioned his environmentalist credentials. Still others were appalled, in the wake of the 1994 Republican ascendancy in the state and national legislatures, that his words would give succor to the then-looming conservative and antienvironmental backlash embodied in the self-styled "Wise Use" movement. In a careful formulation of this challenge, Cohen inquired how "is it possible to offer a constructive critique of environmentalism…especially its 'save the wilderness' version, without damaging valuable parts of the movement, and without offering an argument largely usable by the opponents of environmentalism who are motivated by narrow economic gain?" When it comes to wilderness, one must choose sides and select one's words with care.[221]

But that timorousness begs more questions: Why must environmental historians operate under a form of prior restraint, and why must we take into account how some people might read our work and perhaps draw from our conclusions their own ends? That for some self-censorship is a preferred alternative to searching evaluations of the cultural wellsprings of our fascination with wilderness is precisely why the debates over the significance of old-growth forests or the sanctity of national parks or the intricacy of riparian ecosystems will continue to energize and trouble environmental historiography well into the twenty-first century.

Just as tumultuous have been (and will remain) efforts to assess the interior landscape of those engaged in environmental matters. Part of what makes biography such a tangled and difficult enterprise, as Richard White recognized, is that its practitioners must meld with their subjects (how else know them?) and be a detached observer of those particular lives (how else evaluate their import?). It only adds to the tension should the biographers themselves hold "strong environmentalist sympathies": the resulting books may be "more deeply felt than most conventional academic works." Although "this identification with their subjects gives the books real power, it also presents significant pitfalls." One of these is the manner in which writers then reconstruct the cultural and physical worlds in which their subjects operated. "Nature is at once a physical setting where living beings exist in complex relationships with each other, and a human invention," White noted. "Humans create a shifting set of cultural concepts about the physical world and identify these concepts as nature. When they act, humans do so on the basis of these cultural formulations, but their actions redound on the physical world." Identifying and analyzing these variables immensely complicate the biographical impulse.[222]

The work must be done nonetheless, and a sterling example of the benefits that can flow from rigorous contextual analysis is Steven J. Holmes's *The Young John Muir* (1999). Debunking many of the myths that Muir wrapped around himself and peeling away the layers of legend in which subsequent writers had swaddled him, Holmes places these "traditional images of Muir" in what he denotes as Muir's "environmental biography." This term, and the enterprise it launches, leads him to offer an oft-nuanced account of Muir's "patterns of relationship with the specific environments—natural, domestic, and built—in which he lived and moved...over the course of his lifetime." Patterns that reflect an essential reciprocity—in Holmes's recounting, Muir is as much an active agent, whose "apprehensions of the larger forces and powers of the universe" allow him to "respond creatively and spontaneously to each encounter with a new environment," since he is a reactive figure. Nature's power to require or evoke "new patterns of imagery, feeling, and behavior on the part of humans" constitutes an "independent actor" in the life of a man an earlier biographer had anointed "Son of the Wilderness."[223]

The theoretical underpinning to Holmes's claims depends heavily on his reading of Jean Piaget, Erik Erickson, and other developmental psychologists. Their emphasis on identity formation, when combined with arguments gleaned from environmental psychology, are at the heart of an "object relations approach" that Holmes develops to define Muir's evolving emotional response to nature. The affective bonds he felt for Yosemite, for example, allowed Muir to assimilate this environment "into himself," however incompletely realized. "Despite his own attempts to portray them as static, Muir's relationships to the natural world were…as complex and changing as his relationships with family, friends, human society, and his own self-image—all of which were themselves intimately interrelated with each other, in dynamic and unexpected ways, over the course of his entire life."[224]

Less indebted to psychological models, Curt Meine nevertheless reaches similar conclusions about the demands and requirements of environmental biography. In his essay "Wallace Stegner: Geobiographer," he notes that "our lives are never entirely 'interior' or 'exterior' but always a dynamic interpenetration of both," which leads him to urge greater attention to "the environmental context of biographical studies—the places a subject shapes and is shaped by." This approach has become all the more necessary, for as "psychology has revolutionized our perception of the inner world, advances in the natural sciences have revolutionized our perception of the world around us. Biographies—of some subjects at least—can never be the same." The changes are already evident in a clutch of recent life studies of naturalists and scientists; with varying degrees of success, works on the Bartrams (father and son), George Perkins Marsh, and Rachel Carson have paid attention to the reciprocal exchange between individual and landscape, and the imagination that binds them together. "We selectively perceive what we are accustomed to seeing," once wrote David Lowenthal, who has updated and revised his much-lauded 1958 biography of Marsh. "[F]eatures and patterns in the landscape make sense to us because we share a history with them"; without those associations, without "the past as tangible or remembered evidence[,] we could not function."[225]

That is equally true for those environmental historians more comfortable with the methodologies of cultural, urban, or political analysis, or the ecological or geosciences. Their pursuit of intellectual depth, framed by interdisciplinary explorations and a grudging acceptance of the chaotic nature of nature (human and otherwise), should lead us to thicken our analyses even as we doubt our capacity to understand fully what we are analyzing. That is as it ought to be: "Humans may *think* what they want," Richard White concluded, but "they cannot always *do* what they want, and not all they do turns out as planned."[226]

So Sidney Lanier would confirm: in a split second, while frolicking in Nature's sunny beneficence, he was caught out in the open by an ominous and lowering sky, and left baffled, buffeted, and very wet.

# Acknowledgments

There is no other way of saying it: I have been utterly spoiled by the magnificent and sustained support I have received during my twenty-three years at Trinity University. My colleagues in the history department have been generous with their time and insights; the Joullian Fund, a departmental research endowment, has been a godsend; the university administration has been consistently encouraging and engaged; the staff—most notably senior secretary Eunice Herrington—have made it possible for me to develop as a teacher and scholar; and my many students have challenged me to think more critically and to write more fluently. What's not to like?

Praise is due to a host of others. At the Forest History Society, I have benefited again and again from the energetic leadership of President Steven Anderson and his predecessor, Harold K. (Pete) Steen; good friends, they have also been tough critics. Librarian Cheryl Oakes has been a tremendous resource (I've yet to stump her!); Andrea Anderson runs a tight ship, and all who work there are in her debt for that; and Sally Atwater did a wonderful job copy-editing the manuscript, as did James G. Lewis, recently hired as staff historian, who is a valued colleague: we've shared notes and ideas for years, and I am grateful for his permission to republish here an article we coauthored.

I owe great thanks to the many editors who have thought my ideas worthy of publication, and who—to a man and woman—have helped me write stronger sentences and argue more effectively. At the *Journal of Forestry*, former Publications Director Rebecca Staebler and Managing Editor Fran Pflieger were astonishingly supportive. So was Steven Anderson (again) through his work for *Forest History Today* and *Environmental History*, the latter of which is a joint publication venture of the Forest History Society and American Society of Environmental History. I am grateful for their permission to republish from their pages, and thank for the same reason the relevant editors at *Arnoldia, Pacific Historical Review, The Pinchot Letter*, and Krieger Publishing Company.

My family has been, as always, a source of nurture, and I appreciate their willingness to put up with the stacks of books and papers that like flotsam and jetsam are cast up on the dining room table. Despite that evidence of my presence, they've had good cause to wonder if I don't have a second home in Milford, Pennsylvania, so often have I been to Gifford Pinchot's former manse in the Delaware River Valley. Really, it is all Eddie's fault. That's Edgar B. Brannon, who retired as director of Grey Towers National Historic Landmark in 2004. He has been so important to the development of my appreciation for environmental dilemmas, past and present, that I happily and with great affection dedicate *Ground Work* to him. Warm, wise, and very funny, a skilled administrator and an ebullient soul, he has taught me much about how to live, and live well.

## Credits

Chapter 1, "Pivot Point," was originally published as "The Pivotal Decade: American Forestry in the 1870s" in the *Journal of Forestry* in November 2000, and it is reprinted with the kind permission of the Society of American Foresters.

Chapter 2, "Why *Garden and Forest* Mattered," was originally published as "A High-Grade Paper: *Garden and Forest* and Nineteenth-Century American Forestry" in *Arnoldia* (volume 60, number 2) in Fall 2000 and is reprinted with the kind permission of the Arnold Arboretum.

Chapter 3, "Wooden Politics," was originally published as "Wooden Politics: Bernhard Fernow and the Quest for a National Forest Policy, 1876–1898," in Harold K. Steen, ed., *The Origins of the National Forests* (Durham: Forest History Society, 1992). It appears here with the kind permission of the Forest History Society.

Chapter 4, "What Really Happened in the Rainier Grand Hotel?" was originally published in the *Journal of American History* in March 2000, and is reprinted with the kind permission of the Organization of American Historians.

Chapter 5, "Sawdust Memories," was originally published as "Sawdust Memories: Pinchot and the Making of Forest Service History" in the *Journal of Forestry* in February 1994. It appears here with the kind permission of the Society of American Foresters.

Chapter 6, "Eminent Domain," was originally published as "Eminent Domain: B. L. Wiggins, Forestry, and the New South at Sewanee" in *Forest History Today* in 2004, and is reprinted with the kind permission of the Forest History Society.

Chapter 7, "Groves of Academe," coauthored with James G. Lewis, was originally published as "A Contested Past: Forestry Education in the United States, 1898–1998" in the *Journal of Forestry* in September 1999; it is reprinted with the kind permission of the Society of American Foresters.

Chapter 8, "Grazing Arizona," was originally published in *Forest History Today* in Fall 1999 and is reprinted with the kind permission of the Forest History Society.

Chapter 9, "Back to the Garden," was originally published in *Forest History Today* in Spring 2000. It is reprinted with the kind permission of the Forest History Society.

Chapter 10, "Logjam," was originally published as "Did the Forest Congress Break the Logjam?" in the *Pinchot Letter* in Spring 1996 and is reprinted with the kind permission of the Pinchot Institute of Conservation.

Chapter 11, "Snapshot, 1897," was originally published as "Char Miller on Gifford Pinchot, Photographer" in *Environmental History* in April 2003 and is reprinted with the kind permission of the American Society for Environmental History and the Forest History Society.

Chapter 12, "Green Screen," was originally published as "Green Screen: Projections of American Environmental Culture," in Martin Melosi and Philip Scarpino, eds., *Public History and the Environment* (Chicago: Kreiger Publishers, 2004) and appears here with the kind permission of Kreiger Publishers.

The Afterword, "An Open Field," was originally published in *Pacific Historical Review* in February 2001 and is reprinted with the kind permission of its editors.

# Endnotes

❧

### A Greater Good: An Introduction

1. "Economic Geography," Editor's note, *Economic Geography* 1, no. 1 (March 1925): n.p.; William B. Greeley, "The Relation of Geography to Timber Supply," ibid., 1.
2. Greeley, "Relation of Geography to Timber Supply," 1.
3. Ibid., 1–2.
4. Ibid., 3–5; Carl Bode, ed., *The Portable Thoreau*, rev. ed. (New York: Penguin, 1982), 78–79.
5. Greeley, "Relation of Geography to Timber Supply," 8–11.
6. Ibid., 12–14.
7. James C. Scott, *Seeing Like a State: How Certain Schemes to Improve the Human Condition Have Failed* (New Haven: Yale University Press, 1998), 14.
8. John R. Reiger, *American Sportsmen and the Origins of American Conservation*, 3rd ed. (Corvallis: Oregon State University Press, 2001), 6.

### Chapter One: Pivot Point

9. John W. Collier, Ward Shepard, and Robert Marshall, "The Indians and Their Lands," *Journal of Forestry* 31, no.8 (December 1933): 909.
10. Andrew Denny Rodgers III, *Bernhard Eduard Fernow: A Story of North American Forestry* (1951; reprint, Durham: Forest History Society, 1991), 17.
11. Daniel T. Rodgers, *Atlantic Crossings: Social Politics in a Progressive Age* (Cambridge: The Belknap Press of Harvard University Press, 1998), 3–4.
12. Ibid., 421.
13. Donald J. Pisani, "Forests and Conservation, 1865–1890," in *American Forests: Nature, Culture, and Politics*, ed. Char Miller, 18 (Lawrence: University Press of Kansas, 1997).
14. Harold K. Steen, *The U.S. Forest Service: A History* (Seattle: University of Washington Press, 1976), 9.
15. George Perkins Marsh, *Man and Nature: The Earth as Modified by Human Action* (New York: Charles Scribner, 1864), 27–80.
16. Steen, *U.S. Forest Service*, 13; see also Harold Clepper, ed., *Origins of American Conservation* (New York: John Wiley & Sons, 1966).
17. Michael Williams, *Americans and Their Forests: A Historical Geography* (Cambridge: Cambridge University Press, 1989), 376–77; C. S. Sargent, *Report on the Forests of North America (Exclusive of Mexico), Vol. 9 of the Tenth Census of the United States* (Washington, DC: Government Printing Office, 1880).
18. Franklin B. Hough, *Report upon Forestry* (Washington, DC: Government Printing Office, 1878), 6–9; Steen, *U.S. Forest Service*, 17.
19. F. P. Baker, "Forestry," in *Reports of the United States Commissioners to the Paris Universal Exposition, 1878*, vol. 3 (Washington, DC: Government Printing Office, 1878), 391.
20. Rodgers, *Atlantic Crossings*, 8–9; Baker, "Forestry," 393–94.
21. Baker, "Forestry," 398.
22. Ibid., 423.
23. Ibid.
24. Steen, *U.S. Forest Service*, 123; Rodgers, *Fernow*, 12; Baker, "Forestry," 423.
25. Rodgers, *Fernow*, 93.
26. Gifford Pinchot, *Breaking New Ground*, 4th ed. (Washington, DC: Island Press, 1998), 29.

### Chapter Two: Why *Garden and Forest* Mattered

27. S. B. Sutton, *Charles Sprague Sargent and the Arnold Arboretum* (Cambridge: Harvard University Press, 1970), 131–33.
28. Ibid., 133.
29. Scott, *Seeing Like a State*, 1–22.

30. *Marsh, Man and Nature*, passim; Pisani, "Forests and Conservation, 1865–1890," 15–34; Rogers, *Atlantic Crossings*, passim.
31. *Garden and Forest*, October 6, 1897, 397; August 7, 1895, 319; Carl A. Schenck, "Private and Public Forestry," *Garden and Forest*, June 16, 1897, 232–33; June 23, 1897, 242–43; June 30, 1897, 252; July 7, 1897, 262; Bernhard Fernow, "European Forest Management," *Garden and Forest*, November 11, 1888, 454–55; "Why We Need Skilled Foresters," *Garden and Forest*, April 5, 1895, 131; Gifford Pinchot, "The Forest," *Garden and Forest*, July 30, 1890, 374; August 6, 1890, 386; Pinchot, "Forest-Policy Abroad," *Garden and Forest*, January 7, 1891, 8–9.
32. Gifford Pinchot, "The Need for Forest Schools in America," *Garden and Forest*, July 24, 1895, 298; one aspect of the professionalization of forestry on which Sargent and Pinchot publicly disagreed was the use of military officers as forest guards on the nation's forests. Sargent favored training military officers in the principles of forestry, while Pinchot believed that only a professional civilian service was appropriate; *Garden and Forest*, December 3, 1890, 581; January 1, 1891, 9; January 7, 1891, 9; January 21, 1891, 34–35. As intense as their disagreements would become, they were convinced that the federal government must patrol these valuable public lands; their care could not be left to the states or corporations.

**Chapter 3: Wooden Politics**

33. Bernhard Fernow, "Birth of a Forest Policy," Bernhard Fernow Papers, Cornell University.
34. Ibid.; this anecdote fits within a larger pattern of the discussion of the professionalization of American science in the late nineteenth century: A. Hunter Dupree, *Science in the Federal Government: A History of Policies and Activities* (Baltimore: Johns Hopkins University Press, 1986), 157–169, 239–244; Ronald C. Tobey, *The American Ideology of National Science, 1919–1930* (Pittsburgh: University of Pittsburgh Press, 1971), 3–18; Charles E. Rosenberg, "Science and American Social Thought," in David D. Van Tassel and Michael G. Hall, eds., *Science and Society in the United States*, 135–62 (Homewood, IL: Dorsey Press, 1966); Carrol W. Pursell, Jr., "Science and Government Agencies," 223–50, in ibid.; George H. Daniels, ed., *Nineteenth-Century American Science: A Reappraisal* (Evanston: Northwestern University Press, 1972).
35. Bernhard Fernow, "Government Forestry, 1876–1898," (Washington, DC: Government Printing Office, 1899), 7; he added to his extraordinary publication record, and by the time of his death in 1923, he had published more than 250 articles and bulletins, as well as several books—a productivity that might just put contemporary, computer-aided scholarship to shame.
36. Ben Twight, "Bernhard Fernow and Prussian Forestry in America," *Journal of Forestry*, 88, no.2 (February 1990): 21–25; Char Miller, "The Prussians are Coming! The Prussians are Coming!: Bernhard Fernow and the Roots of the USDA Forest Service," *Journal of Forestry*, 89, no. 3 (March 1991): 23–27, 42.
37. Rodgers, *Fernow*, 14–17; on Germany's extensive influence on American scientific culture and the belief, which Fernow also embraced, that science had a social responsibility, see: Rosenberg, "Science and American Social Thought," 145–58; Tobey, *American Ideology of National Science*, 3–18; Dupree, *Science in the Federal Government*, 221 and passim.
38. Bernhard Fernow to Gifford Pinchot, May 10, 1888, Gifford Pinchot Papers, Library of Congress.
39. Fernow, "Government Forestry," 6–9; Rodgers, *Fernow*, Chapters 2, 3; Steen, *U.S. Forest Service*, 22–46.
40. Fernow to J. G. Kern, March 10, 1887; Fernow to Kinney, April 6, 1887, quoted in Steen, *U.S. Forest Service*, 40.
41. Samuel P. Hays, *Conservation and the Gospel of Efficiency: The Progressive Conservation Movement, 1890–1920* (Cambridge: Harvard University Press, 1959); Dupree, *Science in the Federal Government*; Donald Worster, *American Environmentalism: The Formative Period, 1860–1910* (New York: John Wiley & Sons, Inc., 1973), 73, all suggest that Fernow accomplished little during his tenure as division chief, believing, as Worster has put it, that all he was able to do was keep his "desk tidy." Only Worster has read Fernow's political commentary with some sense of its implications for a new form of governance.
42. Fernow, "Forest Policies and Forest Management in Germany and British India," House Document No. 181, 55th Congress, 3rd Session; Bernhard E. Fernow, *The Economics of Forestry* (New York: Thomas Y. Crowell, 1902), 330.
43. Fernow, "Forest Policies and Forest Management in Germany and British India"; Fernow, *Economics of Forestry*, Chapters 9, 10; Bernhard E. Fernow, "What Is Forestry?" *Bulletin No. 5* (Washington, DC: Government Printing Office, 1891), 7–31; Bernhard E. Fernow, "Economic Conditions Antagonistic to Conservative Forest Policy," *Proceedings of the American Association for the Advancement of Science, 1897*, 329–35.

44. Fernow, "Forest Policies and Forest Management," 238–39.
45. Ibid.
46. For a discussion of Americans' reactions to German forestry, see Miller, "The Prussians Are Coming," 23–27, 42; Fernow, "Forest Policies and Forest Management in Germany and British India," 239.
47. Ibid; Fernow, "What Is Forestry?" 14–15. Fernow's emphasis on the historical validity of German methods was overemphasized. As he recognized, European forestry, or at least his idealization of it, had not been fully realized either in Europe generally or in his beloved Germany. What his historical investigations revealed, in fact, was that the United States was not as far behind the Europeans as he had implied. Most countries had not begun a serious program of rational management until the nineteenth century, until after Napoleon. Indeed, many had only begun to buy up large tracts of land after midcentury, and some not until the 1890s. The German states were more advanced than most, but even here Fernow acknowledged that more than half of German forests were owned by private interests, and the majority of these did not employ systematic and rational forestry. Germany was thus a model for America by its successes *and* its failures. See Fernow, "Forest Policy and Forest Management in Germany and British India," passim; Fernow, "Economic Conditions Antagonistic to Conservative Forest Policy," 330; Twight, "Bernhard Fernow," 21–25; Miller, "The Prussians Are Coming," 23–27, 42.
48. Bernhard E. Fernow, "The Providential Functions of Government with Special Reference to Natural Resources," *Science*, August 30, 1895: 252–54.
49. Ibid., 253–54.
50. Ibid., 255. In this set of arguments, Fernow was indebted to the insights of Frank Lester Ward, and especially to his *Dynamic Sociology* and *Psychic Factors of Civilization*, which he cited in his text. Fernow was thus allying himself with one of the progenitors of twentieth-century liberalism.
51. Ibid., 255–57.
52. Ibid., 256–57; like other scientists in this debate, Fernow tended to exaggerate the extent to which laissez-faire policies guided the American polity: Howard S. Miller, "The Political Economy of Science," in *Nineteenth-Century American Science* (see note 34), 99–114.
53. Fernow, "The Providential Functions of Government with Special Reference to Natural Resources," 262–64.
54. Bernhard E. Fernow, "Address on Forestry," *The Forester*, February 1, 1897, 22–28; Rosenberg, "Science and American Social Thought," 154–58; see also, Hays, *Conservation and the Gospel of Efficiency*; Dupree, *Science in the Federal Government*.
55. Fernow, "Address on Forestry," 26–28; Rodgers, Fernow, 232–41; Steen, *U.S. Forest Service*, 37–42.
56. Pinchot, Diaries, June–December, 1895, Pinchot Papers; Fernow, "The Birth of a Forest Policy"; Fernow to H. H. Chapman, January 4, 1912, quoted in Rodgers, *Fernow*, 220, indicates that Fernow wrote the letter for the secretary of Interior but did so grudgingly, to "satisfy the parties which held with Mr. Pinchot." Fernow would later take credit for initiating the letter and thus creating the commission; see Steen, *U.S. Forest Service*, 30–31.
57. Rodgers, *Fernow*, 209–10, 220–25; Pinchot, *Breaking New Ground*, passim; Fernow to Abbott Kinney, October 9, 1896, quoted in Steen, *U.S. Forest Service*, 32; Fernow, "Address on Forestry," 25, on preservationists; Char Miller, *Gifford Pinchot and the Making of Modern Environmentalism* (Washington, DC: Island Press, 2001), 125–36.
58. Rodgers, *Fernow*, 227, argues that Fernow left Washington having completed "the job on which [he] had set his heart." That is only partly true, for Rodgers provides even better evidence of Fernow's disgruntlement, his belief that at Cornell he would gain the respect and support so missing during his Washington years; Fernow to Henri Joly, March 21, 1896, quoted in Rodgers, *Fernow*, 232. See also: Miller, "The Prussians Are Coming," 23–27, 42; Hays, *Conservation and the Gospel of Efficiency*, 29–35; Dupree, *Science in the Federal Government*, 239–46.
59. Fernow, "The New York State College of Forestry," *Science*, October 14, 1898, 494–501; Rodgers, *Fernow*, Chapter 6; Filbert Roth to Gifford Pinchot, March 10, 1900, Pinchot Papers; Filbert Roth, "Great Teacher of Forestry Retires" *American Forestry* 26, no. 316 (April 1920): 209–12.

### Chapter 4: What Really Happened in the Rainier Grand Hotel?
60. Linnie Marsh Wolfe, *Son of the Wilderness: The Life of John Muir* (Boston: Houghton, Mifflin, 1945), 275–76.
61. See Pinchot, *Breaking New Ground*, 105–109; Steen, *U.S. Forest Service*, 30–34.
62. For a more sophisticated analysis of the sources of the later and intense fighting between the USDA Forest Service and the USDI Park Service, see Hal K. Rothman, "'A Regular Ding-Dong

Fight': The Dynamics of Park Service-Forest Service Controversy during the 1920s and 1930s,"
in *American Forests* (see note 12), 109–24.

63. Lawrence Rakestraw, "Sheepgrazing in the Cascades: Muir v. Minto," *Pacific Historical Review*
27, no. 1 (January 1958): 381–82; note that Pinchot did not even warrant a place in the title
of this important article on the sheep-grazing controversy in Oregon.

64. Michael Smith, *Pacific Visions: California Scientists and the Environment* (New Haven: Yale University
Press, 1987), 163.

65. Steen, *U.S. Forest Service*, 32–34; Pinchot, *Breaking New Ground*, 106.

66. Other recountings include Thurman Wilkins, *John Muir: Apostle of Nature* (Norman: University
of Oklahoma Press, 1995), 201; Frederick Turner, *Rediscovering America: John Muir in His Times
and Ours* (New York: Viking, 1985), 312.

67. *The Wilderness Idea*, Florentine Films, 1989.

### Chapter 5: Sawdust Memories

68. Henry David Thoreau, "A Winter's Walk," in Bode, *The Portable Thoreau*, 66.

69. Pinchot, *Breaking New Ground*, 137.

70. Ibid., 16–22; Gifford Pinchot to Dietrich Brandis, August 21, 1893, Pinchot Papers.

71. Gifford Pinchot, *Biltmore Forest* (Chicago: R. R. Donnelly & Sons, 1893), passim.

72. Gifford Pinchot to Mr. Wetmore, March 21, 1893, Pinchot Papers; William Schlich to Carl Schenck,
January 13, 1897, Carl Schenck Papers, University Archives, North Carolina State University.

73. Steen, *U.S. Forest Service*, 60–64.

74. Gifford Pinchot, "The Lines Are Drawn," *Journal of Forestry* 17, no. 8 (December 1919): 899–900;
"Where We Stand," *Journal of Forestry* 18, no. 5 (May 1920): 441–47.

75. Edgar R. Nixon, *Franklin D. Roosevelt and Conservation, 1911–1945* (New York: Arno Press, 1972),
vol. 1, 129–32, 134.

76. Steen, *U.S. Forest Service*, 199.

77. Richard Polenberg, *Reorganizing Roosevelt's Government: The Controversy over Executive Reorganization,
1936–1939* (Cambridge: Harvard University Press, 1966), 104–105; Pinchot, "Notes of G. P.'s
Talk with Secretary Fall, July 29, 1921," Pinchot Papers; Harry Slattery to Henry S. Graves,
December 21, 1920, Harry Slattery Papers, Special Collections Library, Duke University.

78. Polenberg, *Reorganizing Roosevelt's Government*, 101–05.

79. Ibid., 112–22; T. H. Watkins, *Righteous Pilgrim: The Life and Times of Harold Ickes, 1875–1952*
(New York: Henry Holt, 1990), 561.

80. Bode, *The Portable Thoreau*, 66.

### Chapter 6: Eminent Domain

81. Arthur Ben Chitty, "Leases at Sewanee: Founders' Concept of the University Domain," *The
Sewanee Alumni News*, August 15, 1953, 16.

82. Theodore Roosevelt, "The Forest in the Life of the Nation," *Proceedings of the American Forest
Congress* (Washington, DC: H. M. Suter Publishing Co., 1905), 4–5; 10–11.

83. Ibid., 11.

84. Ibid.

85. James Wilson to Gifford Pinchot, February 1, 1905, Pinchot Papers.

86. B. Lawton Wiggins, "The Attitude of Educational Institutions toward Forestry," in *Proceedings
of the American Forest Congress* (see note 82), 30–31.

87. John Foley, *Conservative Lumbering at Sewanee, Tennessee* (Washington, DC: Government Printing
Office, 1903), 8–10; Arthur Ben Chitty, "Sewanee Now and Then," *Tennessee Historical Quarterly*
34, no. 4 (Winter 1979): 383–84.

88. Joseph H. Parks, *General Leonidas Polk, C.S.A.: The Fighting Bishop* (Baton Rouge: Louisiana State
University Press, 1962), 117–24.

89. Ibid., 124; 138–39.

90. Ibid., 131, n. 22.

91. Ibid., 163–64; Edgar Legare Pennington, "The Battle at Sewanee," *Tennessee Historical Quarterly*
9, no. 3 (1950): 218–19.

92. Parks, *Leonidas Polk, C.S.A.*, 170, 381–86; Pennington, "The Battle at Sewanee," 230–31.

93. Charles Reagan Wilson, "Bishop Thomas Frank Gailor: Celebrant of Southern Tradition,"
*Tennessee Historical Quarterly* 38, no. 3 (Fall 1979): 325–26.

94. William Alexander Percy, *Lanterns on the Levee: Recollections of a Planter's Son* (New York: Alfred
Knopf, 1941), 93–96.

95. Chitty, "Sewanee Then and Now," 383–84; Charles T. Quintard, Diary 1880; and B. Lawton Wiggins to S. McBee, August 23, 1898, quoted in Arthur Ben Chitty to Charles Edward Cheston, August 4, 1970, DuPont Library Archives, University of the South, Sewanee, Tennessee (hereafter DLA).

96. George F. Fairbanks, *History of the University of the South* (Jacksonville, FL: H. & H. B. Drew Co., 1905), 262–63; B. Lawton Wiggins to George R. Fairbanks, July 7, 1905, DLA; Steverson Oden Moffat, "The Gospel of Forestry According to the University of the South," December 1986, 1, DLA; Foley, *Conservative Lumbering at Sewanee, Tennessee*, 7.

97. T. A. Embrey to B. Lawton Wiggins, January 4, 1894, DLA, University of the South.

98. R. M. Dubose to B. Lawton Wiggins, August 30, 1895; September 28, 1896, DLA; Miller, *Gifford Pinchot*, 77–88, 107–12; Sewanee's struggle to regulate depredations through a new managerial and conservationist regime was in line with similar, contemporary efforts in the Adirondacks, Yellowstone, and the Grand Canyon; see Karl Jacoby, *Crimes Against Nature: Squatters, Poachers, Thieves, and the Hidden History of American Conservationism* (Berkeley: University of California Press, 2001). The university's successes were episodic. Between 1900 and 1915, grazing, fire, and timber thievery diminished considerably, only to resurface in the late teens and early twenties. This resurgence was checked again when, in 1923, George A. Garrett was selected to hold the new endowed chair in forestry and serve as university forester; to regain control of the Domain, he initiated a permit system for timber harvesting, used the campus water tower as a fire lookout, established forest patrols, and trained students to help battle fires. A new Franklin County ordinance "prohibiting free ranging of livestock and a reduction of the annual burn to less than one percent," when combined with the elimination of incendiarism, prompted Vice Chancellor Finney to argue in 1926 that "the University forested area…is in better and healthier condition than at any time during the last decade"; *The University of the South Forest: A Demonstration of Practical Applied Forestry for Student Instruction and Regional Use* (Sewanee: University of the South, Forestry Department, 1953), 1–2.

99. Report of the Vice Chancellor, 1897, 63, quoted in Moffat, "The Gospel of Forestry," 2, DLA.

100. B. Lawton Wiggins to S. McBee, August 23, 1898, quoted in Arthur Ben Chitty to Charles Edward Cheston, August 4, 1898, DLA; D. M. Suter to B. Lawton Wiggins, January 1, 1901, DLA: "Mr. Pinchot considers Sewanee a fine place…from a scientific standpoint, and said it would be an excellent place to practice forestry."

101. Proceedings of the Board of Trustees, July 28 to August 3, 1898, 64, DLA; Steen, *U.S. Forest Service*, 54–55.

102. Miller, *Gifford Pinchot*, 112–17; Proceedings of the Board of Trustees, University of the South, July 27 to August 3, 1899, 95–97, DLA; Carl A. Schenck to B. Lawton Wiggins, February 25, 1899; July 18, 1899; August 8, 1899; December 22, 1899; December 27, 1899; March 23, 1900, DLA.

103. Proceedings of the Board of Trustees, University of the South, July 27 to August 3, 1900, 91–92, DLA.

104. Carl A. Schenck, *Birth of Forestry in America* (Santa Cruz, CA: Forest History Society and the Appalachian Consortium, 1955), 81; Foley, *Conservative Lumbering at Sewanee, Tennessee*, 7–8.

105. Foley, *Conservative Lumbering at Sewanee, Tennessee*, 27–33.

106. Wiggins, "The Attitude of Educational Institutions Toward Forestry," 33, 36.

107. Ibid., 39.

108. Paul M. Gaston, *The New South Creed: A Study in Southern Mythmaking* (New York: Alfred A. Knopf, 1970), 7–8, 68–79.

109. Gifford Pinchot to B. Lawton Wiggins, July 26, 1899, DLA.

110. Benjamin Lawton Wiggins, "An Appalachian Forest Reserve and the South," *Forestry and Irrigation* (September 1905): 416, 421.

111. Biographical Sketch of Vice-Chancellor Benjamin Lawton Wiggins, DLA; Wiggins, "The Attitude of Educational Institutions Towards Forestry," 30.

## Chapter 7: Groves of Academe

112. J. W. Giltmier, "American Forestry Education: Conceived in Controversy, Building a Profession" (Washington, DC: Pinchot Institute for Conservation, 1996), 13.

113. Ibid., 8–15.

114. Pinchot to J. W. Pinchot, February 18 and August 31, 1890, Pinchot Papers.

115. Pinchot, Diary, May 1, 1890, Pinchot Papers.

116. Pinchot, *Breaking New Ground*, 72; Schenck, *Birth of Forestry in America*, 3–4.

117. Schenck, *Birth of Forestry in America*, 118.
118. R. A. Skok, "Forestry Education in the United States," in P. MacDonald and J. Lassoie, eds., *The Literature of Forestry and Agroforestry*, 171 (Ithaca: Cornell University Press, 1996).
119. Ralph Hosmer, "The Progress of Education in Forestry in the United States," *Commonwealth Forestry Review* 2 (1923): 88.
120. Biographical Notes of Henry Graves, Henry Graves Papers, Yale University; Pinchot, *Breaking New Ground*, 152.
121. Hosmer, "Progress of Education in Forestry," 88–89.
122. Ibid.
123. Skok, "Forestry Education in the United States," 173; Henry Graves, "Some Considerations of Policy in Forest Education," *Journal of Forestry* 26, no. 4 (April 1928): 430–55.
124. Henry Graves and Cedric H. Guise, *Forest Education* (New Haven: Yale University Press, 1932), 99–104; H. H. Chapman, "Making the Society of American Foresters a Professional Organization," *Journal of Forestry* 32, no.2 (April 1934): 503.
125. H. H. Chapman, "Report on the Committee Accrediting Schools of Forestry," *Journal of Forestry* 41, no.3 (March 1943): 225–29; H. T. Gisborne, "Is the Society Broad Enough?" *Journal of Forestry* 41, no.7 (July 1943): 543.
126. Myron Krueger, "The Society of American Foresters and Forestry Education," *Journal of Forestry* 50, no.1 (January 1952): 6–7.
127. Gifford Pinchot, *The Training of a Forester* (Philadelphia: J. B. Lippincott, 1937), 1–2; 8–10; 86–87; see also Char Miller, "Old Growth: A Reconstruction of Gifford Pinchot's *A Training of a Forester*, 1914–1937," *Forest and Conservation History* 38, no.1 (1994): 7–15.
128. Paul W. Hirt, *A Conspiracy of Optimism: Management of the National Forests Since World War Two* (Lincoln: University of Nebraska Press, 1994), 59–60, 128.
129. J. E. DeSteiguer and R. G. Marrifield, "The Impact of the Environmental Era on Forestry Education in North America," *Unasylva* 31 (1979): 21–25
130. D. P. Duncan, "Report of the SAF Committee on Educational Policies," *Journal of Forestry* 73, no.1 (January 1975): 60; Duncan, R. A. Skok, and D. P. Richards, "Forestry Education and the Profession's Future," *Journal of Forestry* 87, no.9 (September 1989): 31–37; T. J. Brown and J. P. Lassoie, "What Do Employers Want? Entry Level Competency and Skill Requirements of Foresters," *Journal of Forestry* 96, no. 2 (February 1998): 8–14; V. Alaric Sample, "Meeting the Challenges of Education in Sustainable Forestry," *The Pinchot Letter* Spring 1998: 1–4.

**Chapter 8: Grazing Arizona**

131. Aldo Leopold, *A Sand County Almanac* (New York: Oxford University Press, 1949), 200.
132. Hirt, *A Conspiracy of Optimism*, passim, and *American Forests*, 195–272, develop some of the changes that buffeted the forestry profession in the postwar era; Jerry Franklin, "Toward a New Forestry," *American Forests* (November–December 1989): 37–44; Jack Ward Thomas, "The Instability of Stability," *The Pinchot Letter* Summer 1996: 5–6; this is a condensed version of Jack Ward Thomas, "Stability and Predictability in Federal Forest Management: Some Thoughts from the Chief," *Public Land & Resources Law Review* 17 (1996): 9–23; Michael P. Dombeck, "A Gradual Unfolding of a National Purpose: A Natural Resource Agenda for the 21st Century," letter to all Forest Service personnel, March 2, 1998, U.S. Forest Service Collection, Forest History Society, Durham, NC. Christopher McGory Klyza, *Who Controls Public Lands? Mining, Forestry, and Grazing Politics, 1870–1990* (Chapel Hill: University of North Carolina Press, 1996) challenges the assumption that the changes in language the Forest Service used to describe its behavior amount to a paradigm shift in values; the agency's "privilege of technocratic utilitarianism" remains in force, he asserts. Not until a "new world view reintegrating humanity into nature takes root and flourishes" will we demand and obtain "the institutions that could deal with our growing environmental problems." Internal reform, by that definition, is a contradiction in terms; 107, 160. An even more blunt denunciation of the Forest Service's continued existence appears in Robert H. Nelson, "Rethinking Scientific Management," Discussion paper 99–07, Resources for the Future; Nelson's argument that the agency must change its orientation is predicated on a flawed understanding of its early years; that's when, he says, Pinchot insisted that professional experts and "scientific management" determine all resource allocation decisions free from any political influence, an antidemocratic posture that was sharply at odds with Pinchot's political sensibilities, as is confirmed is his experiences in Arizona, and throughout the West. More nuanced analyses of the Forest Service's late-twentieth-century situation include Randy Barker, "New Forestry in the Next West," in *The Next West: Public Lands, Community, and Economy in the American West*, ed. John A. Baden and Donald Snow, 25–44

(Washington, DC: Island Press, 1997); see also, Hanna J. Cortner and Margaret A. Moote, *The Politics of Ecosystem Management* (Washington, DC: Island Press, 1999).

133. Jack Ward Thomas, "This Time, Our Moment in History, Our Future," speech delivered to the Forest Service Leadership Meeting, Houston, Texas, June 20–23, 1994, 14, U.S. Forest Service Collection, Forest History Society, Durham, NC.

134. On Pinchot's challenges to the Forest Service's behavior and belief, see Char Miller, "Sawdust Memories: Pinchot and the Making of Forest Service History," *Journal of Forestry* 92, no. 2 (February 1994): 8–12.

135. Pinchot, *Breaking New Ground*, 177–78.

136. Ibid., 178–79.

137. Ibid., 180–81.

138. Ibid., 181.

139. Gifford Pinchot, "The Use of the National Forests," U.S. Department of Agriculture Forest Service, 1907, 25–26; Pinchot's use of conservation to open up a dialogue on resource use in Colorado is discussed in Char Miller, "Tapping the Rockies: Resource Exploitation in the Intermountain West," in Hal K. Rothman, ed., *Reopening of the West*, 168–182 (Tucson: University Arizona Press, 1998).

140. Gifford Pinchot, "Grazing in the Forest Reserves," *The Forester*, November 1901, 276; Pinchot, *Breaking New Ground*, 181–82.

141. Thomas, "Stability and Predictability in Federal Forest Management," 14–15; Cortner and Moote, *The Politics of Ecosystem Management*, 50–52; the range and seriousness of the challenges that will arise in the shift to ESM are amply documented in K. Norman Johnson, et.al., "Sustaining the People's Lands: The Committee of Scientists' Report," *Journal of Forestry* 97, no. 5 (May 1999): 6–12; responses to the report may be found in ibid., 13–47.

142. Thomas, "Stability and Instability in Federal Forest Management," 18; Thomas, "Comments of Dr. Jack Ward Thomas…," before the National Capital Chapter, Society of American Foresters, September 14, 1994, 7; Dombeck, "A Gradual Unfolding of a National Purpose," 3, 11.

**Chapter 9: Back to the Garden**

143. Information about Rainforest Crunch® from the website of Community Products, Inc., "The Creation of Ben and Jerry's Ice Cream," http://www.madriver.com/thestore/vtfoods.htm.

144. Pinchot, *Breaking New Ground*, 21–22.

145. Ibid., 48.

146. Gifford Pinchot to James W. Pinchot, February 8, 1893, Pinchot Papers.

147. Ibid.

148. Ibid.

149. E. J. Carlson, "The Cherokee Indian Forest of the Appalachian Region," *Journal of Forestry* 51, no. 9 (September 1953): 628–30.

150. Pinchot, *Breaking New Ground*, 411–12.

151. Ibid., 24–25; 412.

152. Ibid., 412.

153. Jay P. Kinney and William Heritage, *The Office of Indian Affairs: A Career in Forestry*, ed. Elwood R. Maunder and George Thomas Morgan, Jr. (New Haven, Conn.: Forest History Society, 1969), 10–30, 44–46.

154. Ibid., 14–15, 50–56.

155. Felix M. Keesing, "The Menomini Indians of Wisconsin: A Study of Three Centuries of Cultural Contact and Change," *Memoirs of the American Philosophical Society* 10 (1939): 168–70, 172–4, 183–87, 232–34, describes the history of the reservation's lumber business and notes some of the internal conflicts it generated; Patricia K. Ourada, *The Menominee Indians: A History* (Norman: University of Oklahoma Press, 1979), 169–189; Nicholas C. Perloff, *Menominee Drums: Tribal Terminations and Restoration, 1954–1974* (Norman: University of Oklahoma Press, 1982), 43–44; Catherine M. Mater, "Menominee Tribal Enterprises: Sustainable Forestry to Improve Forest Health and Create Jobs," in *The Business of Sustainable Forestry: Case Studies* (Chicago: A Project of the Sustainable Forestry Working Group, The John D. and Catherine T. MacArthur Foundation; Washington, DC: Island Press, 1998) argues that one sign of the Menominee's early commitments to conservative land management is that "Today, [they] remain the only Native American tribe to have their forestlands independently certified as being sustainably managed," 9–10; J. P. Kinney, *Indian Forest and Range: A History of the Administration and Conservation of the Redman's Heritage* (Washington, DC: Forestry Enterprises, 1950), 116–41.

156. J. P. Kinney, *A Continent Lost—A Civilization Won* (Baltimore: Johns Hopkins University Press, 1937); *Indian Forest and Range: A History of the Administration and Conservation of the Redman's Heritage* (Washington, DC: Forestry Enterprises, 1950); *My First Ninety-Five Years* (Hartwick, NY: self-published, 1975), 78–9; Kinney, *Office of Indian Affairs*, 26, 31, 33. Kinney's dispute of contemporary assumptions about the damage that the allotment system wrought flies in the face of then-prevailing (and now-current) historiography; for a concise discussion of the damages of the allotment system, see Donald Pisani, "The Dilemmas of Indian Water Policy, 1887–1928," in *Fluid Arguments: Water in the American West*, ed. Char Miller, Chapter 5 (Tucson: University of Arizona Press, 2001).

157. Kinney, *Office of Indian Affairs*, 69–71, 90–93.

158. Collier, "The Indians and Their Lands," 905–906.

159. Ibid., 909.

160. Ferdinand A. Silcox, "Foresters Must Choose," *Journal of Forestry* 33, no. 3 (March 1935): 198–99; 203; Gifford Pinchot to Raphael Zon, October 3, 1934, Raphael Zon Papers, Minnesota Historical Society, St. Paul.

161. H. H. Chapman, "The Responsibilities of the Profession of Forestry in the Present Situation," *Journal of Forestry* 33, no.3 (March 1935): 204–10; V. Alaric Sample and Roger Sedjo, "Sustainability in Forest Management: An Evolving Concept," *International Advances in Economic Research* 2, no. 2 (1996): 171.

162. "The Forest Stewardship Council: Principles and Criteria for Natural Forest Management," *Journal of Forestry* 93, no. 4 (April 1995): 15; "FSC Principles," *FSC Notes* 1, no. 1 (Summer 1995): 4.

163. William Balée, "Indigenous History and Amazonian Diodiversity," in *Changing Tropical Forests: Historical Perspectives on Today's Challenges in Central and South America*, ed. Harold K. Steen and Richard P. Tucker, 192–95 (Durham: Forest History Society, 1992); Leslie E. Sponsel, "The Environmental History of Amazonia: Natural and Human Disturbances, and the Ecological Transition," ibid., 238–45; Warren Dean, "The Tasks of Latin American Environmental History," ibid., 10–13; Lim Hin Fui, "Aboriginal Communities and the International Trade in Non-Timber Forest Products: The Case of Penisular Malaysia," in *Changing Pacific Forests: Historical Perspectives on the Forest Economy of the Pacific Basin*, ed. John Dargavel and Richard Tucker, 77–88 (Durham: Forest History Society, 1992).

164. Hal K. Rothman, *The Greening of America? Environmentalism in the United States Since 1945* (New York: Harcourt Brace & Company, 1998), xi–xvi; Samuel P. Hays, "From Conservation to Environment: Environmental Politics in the United States Since World War II," in *Out of the Woods: Essays in Environmental History*, ed. Char Miller and Hal K. Rothman, 101–126 (Pittsburgh: University of Pittsburgh Press, 1997).

165. Char Miller, "Grazing Arizona: Public Land Management in the Southwest," *Forest History Today* Fall 1999: 15–19; William deBuys, "St. Francis in the Low Post, or, Sustainable Development and the Nature of Environmental Stories," in *Human/Nature: Biology, Culture, and Environmental History*, ed. John P. Herron and Andrew G. Kirk, 79–90 (Albuquerque: University of New Mexico Press, 1999); see also, Cortner and Moote, *Politics of Ecosystem Management*; for a description of some of the cultural and political tensions evident at the Seventh Forest Congress, see Chapter Ten.

### Chapter 11: Snapshot, 1897

166. Nancy Newhall, ed., *The Daybooks of Edward Weston*, vol. 2 (New York: Aperture, 1973), 239. I am grateful to my Trinity University colleague Patricia Simonite for guiding me to Weston's insight.

167. Pinchot, Diary, July 18, 1897, Pinchot Papers.

168. Pinchot, Diary, August 13–15, 1897, Pinchot Papers.

169. Noting that the "development of valuable mines has been almost wholly wanting" along the western Cascades, Pinchot observed in his 1898 report that Monte Cristo was the exception, and he speculated that "it is by no means impossible that others of similar value may be discovered and worked hereafter." But even the Monte Cristo strike proved less substantial than it appeared to Pinchot in 1897; *Survey of Forest Reserves*, U.S. Geological Survey, 55th Congress, 2nd Session, Document No. 189, 114.

170. Miller, *Gifford Pinchot*, 108; and for this reason, had I known about this photograph I would have inserted it into the biography!

171. Pinchot Diary, August 16, 1897, Pinchot Papers.

### Chapter 12: Green Screen

172. Larry Hott to Char Miller, June 14, 1998. Hott was co-producer of *The Wilderness Idea*, which originally aired on PBS in 1989.

173. William Stott, *Documentary Expression and Thirties America* (New York: Oxford University Press, 1973), 5–17; Jerry Kuehl, "Truth Claims," in *New Challenges for Documentary*, ed. Alan Rosenthal, 103–09 (Berkeley: University of California Press, 1988); this tension is repeatedly discussed in "Reel History," a special issue of *Perspectives*, newsletter of the American Historical Association, April 1999. See Robert Brent Toplin, "Film and History: State of the Union," 1, 8–9; Richard White, "History, the Rugrats, and World Championship Wrestling," 12–13; and Simon Schama, "Shooting Britannia," 21–22, 24.

174. Philip Rosen, "Document and Documentary: On the Persistence of Historical Concepts," in *Theorizing Documentary*, ed. Michael Renov, 75–77 (New York: Routledge, 1993); Jay Rubin quoted in Richard M. Barsam, *Nonfiction Film: A Critical History* (Bloomington: Indiana University Press, 1992), 376.

175. Jack C. Ellis, *The Documentary Idea: A Critical History of English-Language Documentary Film and Video* (Englewood Cliffs: Prentice Hall, 1989), 15–28; Stott, *Documentary Expression*, 9.

176. Ellis, *Documentary Idea*, 64–5; for additional analysis of the differences in Flaherty's and Grierson's films, see Barsam, *Nonfiction Film*, 46–55, 89–94; Stott, *Documentary Expression*, 9–12.

177. Grierson quoted in Ellis, *The Documentary Idea*, 73.

178. Richard Dyer MacCann, *The People's Films: A Political History of U.S. Government Motion Pictures* (New York: Hastings House Publishers, 1973), 79–85.

179. Ibid., 79–80; in response to the political charges leveled against *The Plow*, the last minutes were removed from many of the copies of the controversial film shown in theaters.

180. Stott, *Documentary Expression*, 62; MacCann, *People's Films*, 79–85; Barsam, *Nonfiction Film*, 151–57.

181. Bill McKibben, "Curbing Nature's Paparazzi," reprinted in *Harper's Magazine*, November 1997, 19–21, 24.

182. Ellis, *Documentary Idea*, 26–27; Barsam, *Nonfiction Film*, 43; Alexander Wilson, *The Culture of Nature: North American Landscape from Disney to the Exxon Valdez* (Cambridge MA: Blackwell, 1992), 124–25.

183. Wilson, *Culture of Nature*, 125.

184. Ibid., 126–8; see also, David Guss, *The Language of the Birds: Tales, Texts and Poems of Interspecies Communication* (Berkeley: North Point Press, 1985).

185. "Introduction," Rosenthal, ed., *New Challenges for Documentary*, 426.

186. Donald Watt, "History on the Public Screen: I," in *New Challenges for Documentary* (see note 173), 436, 442.

187. Ibid.; an American counterpart of the Open University is the Center for History in the Media at George Washington University; for a discussion of some of its work and the comparable tensions that have emerged, see James G. Lewis, "History, Lies, and Videotape: Historical Documentaries in the Classroom," *OAH Council of Chairs Newsletter* April and June 1997, 1–5; Toplin, "Film and History," 9, makes mention of similar ventures.

188. Jerry Kuehl, "History on the Public Screen: II," in *New Challenges for Documentary* (see note 184), 444–46; Richard White, after extensive work on the documentary *The West*, concluded that the tension between filmmakers and historians had much to do with the differences in the media in which they presented their work: "Documentaries are primarily a visual medium, and, secondarily, an oral medium. Academic history is primarily a print medium." This striking difference led White to propose to Geoff Ward, lead writer of *The West*, that the kind of writing that worked best in documentaries was "topic sentences." Ward countered, "There are no topic sentences. The pictures are the topic sentences." White, "History, the Rugrats, and World Championship Wrestling," 12.

189. Kuehl, "History on the Public Screen: II," 453.

190. White, "History, the Rugrats, and World Championship Wrestling," 11–12.

191. My work as a talking head on *The Greatest Good* (USDA Forest Service, 2005), a documentary assessing the history of the Forest Service, reinforced these conclusions.

192. Lewis, "History, Lies, and Videotape," 2.

193. Authoritative omniscience may backfire, of course, leading viewers to be suspicious of that which it "asserts most fervently." The documentary form itself may have changed in reaction to challenges to its narrative voice. As one critic has noted, the "emergence of so many recent documentaries built around strings of interviews" may be a "strategic response to the recognition that neither can events speak for themselves nor can a single voice speak with ultimate authority." For a discussion of these and related issues, see Bill Nichols, "The Voice of Documentary," in *New Challenges for Documentary* (see note 173), 54–55; they are also covered in James Roy MacBean, "Two Laws from Australia, One White, One Black," in *New Challenges for Documentary* (see note 184), 210–26.

194. Toplin, "Film and History," 8.

## Chapter 13: Trees, Sprawl, and Urban Politics

195. E. G. McPherson, "Urban Forestry: The Final Frontier?" *Journal of Forestry* 101, no. 3 (April–May 2003): 21, 24.

196. "Urban Ecosystem Analysis: San Antonio, TX, Region" (Washington, DC: *American Forests*, 2002), 1–2.

197. Mike Greenberg, "Tree-Slaughter Shows Contempt," *San Antonio Express-News*, June 15, 1993, 16D; Susie P. Gonzalez, "Landscape ordinance after old trees felled," *San Antonio Express-News*, June 18, 1993, 8D.

198. R. Bush, "Tree rules fail to stop bulldozers," *San Antonio Express-News*, May 5, 1999, 1H; William Pack, "Tree ordinance up for review," *San Antonio Express-News*, December 17, 2000, 7B.

199. H. W. Lawrence, "Changing Forms and Persistent Values: Historical Perspectives on the Urban Forest," in *Urban Forest Landscapes: Integrating Multidisciplinary Perspectives*, ed. G. Bradley, 31 (Seattle: University of Washington Press, 1995).

200. Frederick Law Olmsted, *Journey through Texas, a Saddle-Trip on the Southwestern Frontier* (Austin: University of Texas Press, 1978), 147–48.

201. L. M. Spell, "The Grant and Survey of the City of San Antonio," *Southwestern Historical Quarterly*, 66, no. 1 (1966): 73–89.

202. Ibid., 82–85.

203. Char Miller, *Deep in the Heart of San Antonio: Land and Life in South Texas* (San Antonio: Trinity University Press, 2004).

204. Laura Wimberly, "Establishing Sole-Source Protection: The Edwards Aquifer and the Safe Water Act," in *On the Border: An Environmental History of San Antonio*, ed. Char Miller, 169–181 (Pittsburgh: University of Pittsburgh, 2001); John Donahue and Jon Q. Sanders, "Sitting Down at the Table: Mediation and Resolution of Water Conflicts," in ibid., 182–98.

205. "Master Plan Policies," adopted by City of San Antonio City Council, City of San Antonio, 1997.

206. Greenberg, "Tree-slaughter shows contempt," 16D.

207. Mike Greenberg, "New urbanism a shift from earlier codes," *San Antonio Express-News*, May 18, 2003, 4B.

208. Char Miller and Heywood Sanders, "Parks, Politics, and Patronage," in *On the Border* (see note 216), 83–98.

209. "Charter of the New Urbanism," in *Toward the Livable City*, ed. Emilie Burkhardt, 277–82 (Minneapolis: Milkweed Editions, 2003).

210. Greenberg, "New Urbanism a Shift from Earlier Codes," 4B; Rick Casey, "Will 'Smart-Growth' Win the Battle and Lose the War?" *San Antonio Express-News*, April 7, 2003, 20; Jerry Needham, "Lobbyist Defends Barrage of Lawsuits," *San Antonio Express-News*, June 4, 2004, 1, 2B.

211. An Ordinance Amending Chapter 21 and Chapter 35 of the City Code of San Antonio, City of San Antonio, 2003.

212. Roddy Stinson, "Utility goliath seeks aid in competing with tiny gas company," *San Antonio Express-News*, January 27, 2004, 3A. In May, two local independent school districts petitioned city council to be made exempt from all provisions of the Tree Preservation Ordinance. After a public uproar, their bid was tabled; Karen Adler, "Council leaves tree law alone," *San Antonio Express-News*, May 7, 2004, 1.

213. McPherson, "Urban Forestry: The Final Frontier?" 24.

214. Jerry Needham, "S.A. to tout tree-friendly efforts," *San Antonio Express-News*, September 17, 2003, 1B.

215. Leopold, *Sand County Almanac*, 207, 225; Needham, "S.A. to tout tree-friendly effort," 1B.

## Afterword

216. Sidney Lanier, "San Antonio de Bexar," *Southern Magazine*, XIII, July 1873, 83–99; August 1873, 138–152.

217. Richard White, "American Environmental History: The Development of a New Historical Field," *Pacific Historical Review* 54, no.3 (August 1985): 297–304.

218. Ibid., 305–306.

219. William Cronon, "The Trouble with Wildernesss; Or, Getting Back to the Wrong Nature," in *Uncommon Ground: Toward Reinventing Nature*, ed. William Cronon, 69–90 (New York: W.W. Norton & Co., 1995); see also *Environmental History* 1, no.1 (January 1996): 7–28; *Out of the Woods* (see note 164), 28–50; a condensed version appeared under the same title in the *New York Times*, August 13, 1995, 42.

220. Cronon, "The Trouble with Wilderness," 20–25.

221. Samuel P. Hays, "The Trouble with Bill Cronon's Wilderness," *Environmental History* 1, no.1 (January 1996): 29–32; Michael P. Cohen, "Resistance to Wilderness," ibid., 33–42; Thomas R. Dunlap, "But What Did You Go Out to Wilderness to See?" ibid., 43–47; less cautious than these published reactions to Cronon's arguments were some of those posted on the American Society of Environmental History listserve (H-Environment: http://www2.h-net.msu.edu/logsearch/). Historian Paul Hirt, for one, argued that "Cronon does not understand the wilderness movement. There is some interesting history in the essay, but when he turns to his critique of wilderness, his argument becomes essentially linguistic. He plays with terms and teases his readers with ironies, but in the end offers little more than a straw dog representation of the movement that is neither informative nor helpful as far as conservation is concerned." To environmentalist Dave Foreman's charge that the worst thing Cronon had done was to give fuel "to the traditional enemies of conservation," Hirt intoned, "Amen." Religious sensibilities are ever present in the politics of wilderness; Paul Hirt, "Cronon vs. Foreman," 3 February 1997.
222. White, "American Environmental History," 305–307.
223. Steven J. Holmes, *The Young Man Muir: An Environmental History* (Madison: University of Wisconsin Press, 1999), 3–9; Wolfe, *Son of the Wilderness.*
224. Holmes, *Young John Muir,* 243, and Appendix C, 265–87.
225. Curt Meine, "Wallace Stegner: Geobiographer," in *Wallace Stegner and the Continental Vision: Essays on Literature, History, and Landscape,* ed. Curt Meine, 121–39 (Washington, DC: Island Press, 1997); Thomas P. Slaughter, *The Natures of John and William Bartram* (New York: Alfred A. Knopf, 1996); Linda Lear, *Rachel Carson: Witness for Nature* (New York: Henry Holt and Company, Inc., 1998); David Lowenthal, *George Perkins Marsh: Prophet for Conservation* (Seattle: University of Washington Press, 2000); David Lowenthal, "Past Time, Present Place: Landscape and Memory," *Geographical Review* 65 (January 1975): 5–6.
226. Donald Worster, "The Ecology of Order and Chaos," reprinted in *Out of the Woods* (see note 164), 3–17; see also the sharp debates over the moral mission, political focus, and historiographical future of environmental history in "A Round Table: Environmental History," *Journal of American History* 76, no.4 (March 1990): 1087–147; White, "American Environmental History," 335.

# Index

**A**

Alamo Forest Partnership, 147
American Association for the
  Advancement of Science, 14
American Forest Congress (1905),
  61–64, 73, 76
American Forest Congress (1996)
  (Seventh), 7, 109, 110–14
American Forest Council, 16
American Forestry Association (*also*
  American Forests), 12, 20, 28, 36,
  54, 139, 147
American Forest Congress. *See*
  American Forestry Association
*American Journal of Forestry*, 15
Arizona Wool Growers Association, 89
Army, U.S. (Cavalry), 36, 43n
Arnold Arboretum, 15, 21, 24, 43n
Attenborough, David, 130
Audubon Society, 111

**B**

Baker, F. P., 16–19, 20
Balée, William, 107–08
Ballinger, Richard, 100
Bartram, John and William, 154
Biltmore Estate, 5, 51, 97, 98, 118
Biltmore Forest School: 5, 71, 75n,
  79–80
  as compared to Yale School of
    Forestry: 78, 79, 80, 81
Bliss, Cornelius, 42, 47, 117, 119
Boppe, Lucien, 51
Brady, Chris, 147
Brandis, Dietrich, 14, 51
Brown, Percy, 75n
Brownlow Committee Report, 56
Bunch, Con, 89

**C**

Carruthers, William, 21
Carson, Rachel, 7, 108, 136, 154
Chapman, H. H., 83–84
Chitty, Arthur Ben, 61
*Circular 21*, 71, 75n
Civil War, U.S., 3, 65, 66, 68, 75
Civilian Conservation Corps, 54
Clarke-McNary Act (1924), 54
Clean Air Act (1970), 7
Clean Water Act (1972), 7
Cleveland, Grover, 29, 35, 36, 42, 117
Cleveland Reserves. *See* Washington's
  Birthday Reserves
Cochran, H. Dean, 85
Colby, William E., 42
Cohen, Ben, 96
Cohen, Michael P., 152
Collier, John, 6, 104–06
Commission of Indian Affairs, 6
Congress of New Urbanism, 144
Conservation movement, 24, 39, 62,
  88
Cortner, Hannah, 94
Cousteau, Jacques, 7, 129, 130–31
Coville, Frederick V., 48, 89
Coville Report, 42, 43n
Cronon, William, 151–52

**D**

Darling, J. N., 57
de Almazán, Juan Pérez, 141
DeBuys, William, 108–09
Department of Agriculture, U.S.: 28,
  38, 62;
  rivalry with Department of Interior:
    55–58, 62, 70, 101

Department of Interior: 38, 84, 89,
    99, 102, 117;
    Indian Office: 6, 99–100;
    rivalry with Department of
        Agriculture: 55–58, 62, 70, 101
Dinosaur National Monument, 108
Disney, Walt, 7, 17, 131
Documentary films:
    history of environmental: 123–37;
    early years: 124–129;
    television era of: 130–33;
    contemporary era of: 132–37
Dombeck, Michael P., 94
*Drifters, The* (Film, 1929), 7, 126


**E**
Earth First!, 110
*Economics of Forestry, The* (1902), 81
Ecosystem Management, 87, 92–94
Edwards Aquifer, 140, 142, 143
Egleston, Nathaniel H., 26, 27, 28, 34
Ehrlich, Paul, 108
Elliott, Stephen, 66
Ellis, Jack, 124
Embrey, T. A., 70
Endangered Species Act, 87, 129
*Environmental History*, 151, 152
Environmental movement, 47, 85,
    108–09, 111, 150, 152
Evans, Brock, 111


**F**
Fairbanks, George R., 68
Fall, Albert, 55–56
Farm Security Administration, 128
Fernow, Bernhard Eduard: 16, 19, 22,
    36, 78, 160n35, 160n37, 161n47;
    early career: 11, 27–28;
    as chief of Division of Forestry: 4–5,
        24, 26–29, 35, 36–37, 160n41,
        161n56, 161n58;
    on forestry: 30–32;
    on American political culture:
        32–35, 161n50;
    as educator: 37–38, 75n, 80–81

Finch, Calvin, 147, 148
Flaherty, Frances, 124
Flaherty, Robert, 7, 124–26, 130, 135
Florentine Films, 121, 134
Foley, John, 71–73
Foreman, Dave, 168n221
*Forest Education* (1932), 83
Forest Management Act (1897), 78,
    80–81
Forest Reserve Act (1891), 29
Forest Service: 1, 53, 100, 101, 105;
    as Division or Bureau of Forestry: 5,
        15, 26, 49, 71, 79, 99;
    Hough as chief of: 15;
    Egleston as chief of: 26–27;
    Fernow as chief of: 4–5, 24, 26–27,
        28–29, 35, 36–37, 160n41,
        161n56, 161n58;
    Pinchot as chief of: 35, 38, 52, 53,
        70–71, 99–100, 163n100;
    establishment of (1905): 5, 15, 38,
        55, 62;
    Graves as chief of: 54;
    Greeley as chief of: 1–4, 54;
    Stuart as chief of: 82;
    Silcox as chief of: 55, 56, 105–06;
    Thomas as chief of: 6, 87–88, 92–94;
    Dombeck as chief of: 94;
    in Agriculture: 38, 62, 70;
    grazing policies of: 49, 87–95;
    transfer threats against: 55–58;
    managerial model of: 82;
    ecosystem management in: 87,
        92–94, 164n132
    timber management: 102;
Forest Stewardship Council, 7, 106–07
Forestry: 1, 74;
    in the United States: 4, 29–30, 32;
    in Europe: 17–19, 20, 23, 25, 97;
    in Germany: 29–32, 97, 113,
        161n47
*Forestry and Irrigation*, 75
Forestry education, history of, 77–86
Forestry movement (U.S.), 24, 25, 27,
    37, 78

**G**

Gailor, Thomas Frank, 67
*Garden and Forest*: 15;
  importance of: 21–25
Gaston, Paul, 74
General Land Office, 29, 46
Geological Survey, U.S., 45n
*Gifford Pinchot and the Making of
  Modern Environmentalism* (2001), 47
Gifford, Sanford, 10, 120
Gisborne, H. T., 83–84
Goodman Lumber Company, 102
Gordon, John, 111
*Grass* (Film, 1925), 130
Graves, Henry S.: 19, 78;
  as chief of Forest Service: 54;
  as dean of Yale Forest School: 53, 81;
  and forestry education: 83, 85
Grazing:
  controversy over: 41–47;
  Forest Service policies on: 49, 87–95
Greeley, William B.: 1, 4
  as associate chief: 54;
  as chief of Forest Service: 1–4, 5, 6,
    54
Greenberg, Mike, 139
Greene, Lorne, 130
Greenfield, Jerry, 96
Grierson, John, 7, 124–26, 130
Grey Towers, 60, 79
Guise, Cedric Hay, 83

**H**

Haines, Sharon, 111
Harriman, Edward Henry, 43n
Hays, Samuel Pinchot, 150, 152
Hermann, Binger, 46, 47
Hetch Hetchy Valley: 121;
  controversy: 47, 48, 134, 136
Heyer, G., 28
Hirt, Paul, 168n221
Hodgson, Telfair, 68
Holmes, J. A., 45n
Holmes, Stephen J., 153–54
Hott, Larry, 121, 124, 136

Hough, Franklin: 13–16, 19, 20, 26;
  as chief of Division of Forestry:
    15–16, 34
Hudson River landscape painting
  style, 10, 120

**I**

Ickes, Harold, 55–58
Indian Forest Service (U.S.), 101, 103,
  104
International Paper, 111

**J**

Johnson, Martin and Osa, 130
*Journal of Forestry*, 53, 54, 84, 104

**K**

Keffer, Charles, 35
Kelley, Steve, 112
Kinney, Jay P.: 100–05, 165n156;
  publications by: 103, 104
Kirby Lumber Company, 75n
Kmart, 139, 143
Krueger, Myron, 84
Kuehl, Jerry, 123, 132–34

**L**

La Follette, Robert, 102
Lanier, Sydney, 149, 154
L'Ecole Nationale Forestière (France),
  24, 51, 78
Leopold, Aldo, 31, 87, 130, 148
Lewis, James G., 77n, 135
Lipscomb, Douglas, 138
Lorentz, Pare, 7, 124, 126–29, 131

**M**

MacCann, Richard, 128
Malin, James, 150
*Man and Nature* (1864), 14, 23
Marsh, George Perkins: 13, 16, 20, 23,
  154;
  and *Man and Nature*: 4, 14, 23
Marshall, Robert, 54, 104, 105
McBee, Silas, 70

McKibben, Bill, 129
McPherson, E. Gregory, 138, 147
Meine, Curt, 154
Menominee Indians, 102–03,
   165n155
Merriam, C. Hart, 34, 46, 49
*Moana* (Film, 1926), 124
Monte Cristo, Washington, 118,
   119–20
Motte, Anne, 94
Mount Baker-Snoqualmie National
   Forest, 118, 119
Muenden, Germany, 28
Muir, John: 36, 130, 136, 153–54;
   dispute with Gifford Pinchot: 39–49
Mulford, Walter, 85

**N**

*Nanook of the North* (Film, 1922), 7,
   124, 125
Nash, Roderick, 150
National Academy of Sciences, 35, 36
National Environmental Policy Act
   (1970), 7, 87
National Forest Commission, 36, 42,
   43, 46, 78, 117, 161n56
National Forest Management Act
   (1976), 87
National Forest System, 31
National Geographic Society, 130,
   131
Native Americans: 2;
   and forestry: 6, 97–100, 104–06
New Deal, 6, 54, 104, 127, 128
New Jersey, forestry in: 34–35
New York State College of Forestry,
   Cornell University, 5, 6, 37–38,
   75n, 80–81, 101, 161n58

**O**

Oberlaender Trust, 12
Olmsted, Frederick Law, 140–41
Open University, 133, 167n187

**P**

Paris Universal Exposition, 16–18
Peak, Howard, 139, 144
Percy, William Alexander, 67
Perkins, Marlin, 130
Pinchot, Amos, 117, 120
Pinchot, Gifford: 10, 20, 22, 78, 136,
   150, 160n32;
   education of: 19, 28–29, 51, 97;
   work at Biltmore: 24, 51, 97–99, 118;
   as consulting forester: 6, 70;
   dispute with John Muir, 39–48;
   as Confidential Forestry Agent: 42,
      117–20, 166n169;
   and National Forest Commission:
      36, 43;
   as chief of Forest Service: 5, 35, 37,
      38, 52, 53, 70–71, 99–100,
      163n100;
   grazing policy of: 90–92, 94–95
   Ballinger controversy: 53, 100;
   on regulating private timberlands:
      54;
   fights Forest Service transfer, 55–58;
   on professional forestry: 25, 36,
      50–51, 54, 81–82, 84–85;
   publications by: 84, 88, 97, 118
Pinchot, James: 10, 39, 120;
   and Yale School of Forestry: 60
Pinchot, Mary, 39
*Plow That Broke the Plains, The* (Film,
   1936), 124, 126–27, 128
Polk, Leonidas, 65–67, 75
Potter, Albert: 89–90;
   as chief of Forest Service grazing:
      92, 93
Pratt, John W., 45
Prussian Forestry Department, 27

**Q**

Quintard, Charles T., 67, 75

**R**

*Rachel Carson's Silent Spring* (Film, 1993), 136
Rakestraw, Lawrence, 42, 43
Reiger, John, 4
*Report on the Forest of North America* (1884), 15
Rescission Bill of 1995 (Salvage Rider), 110, 113
Resettlement Administration, 128
Reynolds, Olivia, 11
*River, The* (Film, 1937), 126, 127–28
Rodgers, Daniel T., 11–12, 14
Roosevelt, Franklin D.: 54, 57, 104, 128;
    and New Deal: 127–28
Roosevelt, Theodore, 5, 61–62, 63–64, 128
Roselle, Mike, 110
Rosen, Philip, 123
Rosenthal, Alan, 132

**S**

Salt River Water Users, 89
Salvage Rider. *See* Rescission Bill of 1995 (Salvage Rider)
Sample, V. Alaric, 106
San Antonio, Texas: 8, 91, 148;
    environmental history of: 138–48;
    and urban sprawl, 8, 138–40;
    settlement of: 139–42;
    modern growth of: 142–48;
    urban forestry in: 144–48
Sargent, Charles Sprague: 15, 22, 43, 160n32;
    and *Garden and Forest*: 15, 21–25;
    and National Forest Commission: 36
Schenck, Carl: 5, 19, 24, 51, 78;
    consults at University of the South: 71–72;
    and Biltmore Forest School: 79–80
Schlich, William, 51
Schurz, Carl, 13, 15
Scott, James C., 4, 23
*Seattle Post-Intelligencer*, 45, 48, 49n, 110

Sedjo, Roger A., 106
Seventh American Forest Congress (1996). *See* American Forest Congress (1996) (Seventh)
Sewanee (University of the South). *See* University of the South
Sewanee Mining Company, 65, 66
Shepard, Ward, 104, 105
Sierra Club, 39, 42n
Silcox, Ferdinand A.: 82;
    as chief of Forest Service: 54, 56, 105–06
*Silent Spring* (1962), 7, 108
Silviculture, teaching of: 23, 80, 81, 84, 85
Skok, Richard, 83
Slattery, Harry, 56
Smith, Michael, 43, 46
Société des Sylviculteurs de France et des Colonies (France), 23
Society of American Foresters: 20, 53, 54;
    and forestry education: 83–84
Society of Range Management, 84
Social Darwinism, 32–33
*Son of the Wilderness*, (1945), 40, 41
Spencer, Herbert, 32–33
*Star Trek V: The Final Frontier* (Film, 1989), 116, 121, 122
Steen, Harold K., 71
Stiles, William A.: 21;
    as editor of *Garden and Forest*: 22, 24
Stott, William, 123, 128–29
Streep, Meryl, 136
Stuart, Robert Y., 82
Sutton, Stephanie B., 21

**T**

Taft, William Howard, 53, 100
Tennessee Valley Authority, 128
Thomas, Jack Ward, 6, 87–88, 92–94
Thoreau, Henry David, 2, 50, 58, 130
Timber Culture Act (1873), 19
Toyota Motor Company, 146
Tracy, Samuel F., 66
*Training of a Forester, The* (1937), 84

Trans-Atlantic exchange of ideas, 4, 11–12, 23, 24, 30, 160n37
Turner, Frederick Jackson, 150

**U**
United States Film Service, 129
University of California (Berkeley) School of Forestry, 85
University of the South (Sewannee): 61–76;
    founding of: 65–67;
    forestry at: 5, 68–75, 85, 86, 163n98
Urban forestry, 8, 73, 144–48
Urban sprawl, 91, 138–40
U.S. Centennial Exposition (1876), 11

**V**
Vanderbilt, George, 6, 79, 80, 97, 98

**W**
Wallace, Henry, 55, 56
Wallinger, R. Scott, 77
Ward, Frank Lester, 34, 161n50
Ward, Geoff, 167n188
Washington's Birthday Reserves, 29, 35, 36, 42, 117
Watershed protection, 91–92
Watt, Donald, 132–34
Webb, Walter Prescott, 150
Weston, Edward, 117

Westvaco Corporation, 77
White, Mark, 144, 146
White, Richard: 8, 149–151, 152, 153, 154;
    on documentary films: 134, 135, 167n188
Wiggins, Benjamin L.: 63, 64, 68;
    supports forestry: 5, 69–76
Wild and Scenic Rivers Act (1968), 87
Wilderness Act (1964), 7, 87, 134
*Wilderness Idea, The* (Film, 1989), 47, 124, 134, 137
Wilson, Alexander, 131–32
Wilson, James, 35, 62–63
"Wise Use" Movement, 152
Wolfe, Linnie Marsh, 41, 43, 47
Worster, Donald, 150, 160n41

**Y**
Yale College, 65, 78
Yale School of Forestry: 5, 6, 53, 60, 73, 85, 111;
    as compared to Biltmore Forest School: 78, 79–80, 81
Yellowstone National Park, 4
Yosemite National Park: 47, 48, 153;
    filming in: 116, 121, 124

**Z**
Zon, Raphael, 54

# About the Author

❧

Char Miller specializes in American environmental, cultural, and urban history at Trinity University in San Antonio, Texas. He served as chair of the History Department from 1998 to 2004, and since 2001 has been Director of Urban Studies.

A Senior Fellow of the Pinchot Institute for Conservation, Miller is a Contributing Writer of the *Texas Observer*, Associate Editor of *Environmental History* and for the *Journal of Forestry*, is on the Editorial Board of the *Pacific Historical Review*, and sits on the Board of Directors of the Forest History Society. A member of the Trustee Advisory Board for the Witte Museum in San Antonio, he has served on the City of San Antonio's Open Space Advisory Board and its Tree Preservation Ordinance Panel.

His books include: *Deep in the Heart of San Antonio: Land & Life in South Texas*; *Gifford Pinchot and the Making of Modern Environmentalism* (award winner); *The Greatest Good: 100 Years of Forestry in America*; *Gifford Pinchot: The Evolution of An American Conservationist*; and *Fathers and Sons: The Bingham Family and the American Mission*. Miller is editor of *Richard Harding Davis, The West From a Car-Window*; *The Atlas of U.S. and Canadian Environmental History*; *On the Border: An Environmental History of San Antonio*; *Fluid Arguments: Five Centuries of Western Water Conflict*; *Water and the Environment: Global Perspectives* (with Mark Cioc and Kate Showers); *Water and the West: A High Country News Reader*; *American Forests: Nature, Culture, and Politics*; and *Out of the Woods: Essays in Environmental History* (With Hal K. Rothman).

# About the Forest History Society

The Forest History Society (FHS) is a 501(c)3 nonprofit educational institution that links the past to the future by identifying, collecting, preserving, interpreting, and disseminating information on the history of interactions between people, forests, and their related resources—timber, water, soil, forage, fish and wildlife, recreation, and scenic or spiritual values. Through programs in research, publication, and education, the Society promotes and rewards scholarship in the fields of forest, conservation, and environmental history while reminding all of us about our important forest heritage.

# Publications of the Forest History Society

❧

If you enjoyed this book, you may be interested in these other books and films from the Forest History Society. To purchase a copy, please contact the Forest History Society at www.foresthistory.org/publications or (919) 682-9319, or visit your local bookseller. We also have nearly 300 oral history interviews for sale. Please see our website for a listing.

## Issues Series

The Forest History Society Issues Series consists of booklets that bring a historical context to today's most pressing issues in forestry and natural resource management.

*Genetically Modified Forests: From Stone Age to Modern Biotechnology*, Rowland D. Burdon and William J. Libby
*Forest Pharmacy: Medicinal Plants in American Forests*, Steven Foster
*American Forests: A History of Resiliency and Recovery*, Douglas W. MacCleery
*Newsprint: Canadian Supply and American Demand*, Thomas R. Roach
*America's Fires: Management on Wildlands and Forests*, Stephen J. Pyne
*Forest Sustainability: The History, the Challenge, the Promise*, Donald W. Floyd
*Canada's Forests: A History*, Ken Drushka

## Publications Relating to Conservation and Forest Service History

*Proceedings of the U.S. Forest Service Centennial Congress: A Collective Commitment to Conservation*, Steven Anderson (ed.), $24.95
*The Forest Service and the Greatest Good: A Centennial History*, James G. Lewis, $30.00 hardcover, $20.00 softcover
*The Chiefs Remember: The Forest Service, 1952–2001*, Harold K. Steen, $29.00 hardcover, $20.00 softcover
*Jack Ward Thomas: The Journals of a Forest Service Chief*, Harold K. Steen (ed.), $30.00
*Pathway to Sustainability: Defining the Bounds on Forest Management*, John Fedkiw, Douglas W. MacCleery, V. Alaric Sample, $8.95
*The U.S. Forest Service: A History, Centennial Edition*, Harold K. Steen, $40.00 hardcover, $25.00 softcover

*The Conservation Diaries of Gifford Pinchot*, Harold K. Steen (ed.), $29.00 hardcover, $19.00 softcover

*Forest and Wildlife Science in America: A History*, Harold K. Steen (ed.), $14.95 softcover

*Millicoma: Biography of a Pacific Northwestern Forest*, Arthur V. Smyth, $12.95 softcover

*Cradle of Forestry in America: The Biltmore Forest School, 1898–1913*, Carl Alwin Schenck, $10.95 softcover

*Forest Service Research: Finding Answers to Conservation's Questions*, Harold K. Steen, $10.95 softcover

*Origins of the National Forests: A Centennial Symposium*, Harold K. Steen (ed.), hardcover $31.95; $16.95 softcover

*Bernhard Eduard Fernow: A Story of North American Forestry*, Andrew Denny Rodgers III, $21.95 hardcover

### Available DVD Videos

*The Greatest Good: A Forest Service Centennial Film* (2005) ($18.00)

*Timber on the Move: A History of Log-Moving Technology*, Vester Dick (1981) ($25.00)

*Up in Flames: A History of Fire Fighting in the Forest*, Vester Dick (1984) ($25.00)